Phase Transitions

for Beginners

Phase Transitions

for Beginners

Sergei M Stishov
Institute for High Pressure Physics
of Russian Academy of Sciences, Russia

NEW JERSEY • LONDON • SINGAPORE • BEIJING • SHANGHAI • HONG KONG • TAIPEI • CHENNAI • TOKYO

Published by

World Scientific Publishing Co. Pte. Ltd.
5 Toh Tuck Link, Singapore 596224
USA office: 27 Warren Street, Suite 401-402, Hackensack, NJ 07601
UK office: 57 Shelton Street, Covent Garden, London WC2H 9HE

British Library Cataloguing-in-Publication Data
A catalogue record for this book is available from the British Library.

PHASE TRANSITIONS FOR BEGINNERS

ISBN 978-981-3274-17-4

For any available supplementary material, please visit
https://www.worldscientific.com/worldscibooks/10.1142/11096#t=suppl

Typeset by Stallion Press
Email: enquiries@stallionpress.com

Printed in Singapore

Foreword

The title of this little book, "Phase transitions for beginners", is in some degree conditional. By "beginners", I mean individuals who have already become acquainted with phase transitions from standard courses in statistical physics and want to continue studying the subject. In the presented material, the reader does not find anything similar to a systematic exposition of the physics of phase transitions. In fact, the whole text is a collection of a kind of scientific "novellas", the content of which, one way or another, is related to the author's interests.

The justification of this publication is the specificity of most modern books on phase transitions, which are mainly devoted to the theory of fluctuations and scale invariance in phase transitions of second order.

This book describes in some detail the physics of melting, tricritical phenomena, and a number of exotic questions, usually not covered in the standard manuals (systems with step potential, "cold" melting, etc.).

In describing the second-order phase transitions, I employ the model of a binary alloy instead of the equivalent and most frequently used Ising model, thereby giving a tribute to the famous Landau article on phase transitions.

Naturally, I could not do without describing the fluctuation effects occurring at the phase transitions. Here I give a simple derivation of the scaling relations, based on dimensional considerations.

In the Appendix, brief information is given on important concepts used in the discussion of phase transitions. Here, too, are my popular article on quantum phase transitions and a kind of essay on some semantic problems that arise at describing phase transitions. Two my recent reviews written with Dr. A. Petrova on physics of the itinerant helical magnet MnSi are included just to show potential readers what is a real experimental work on phase transitions. Finally, a significant part of the illustrations was borrowed from the works of author and his collaborators.

Contents

Chapter 1

Introduction

Phase transitions are transitions between different physical states (phases) of the same substance. Common examples of phase transitions are the ice melting and the water boiling, or the transformation of graphite into diamond at high pressures. Figure 1 schematically shows phase diagrams of matter in temperature-pressure ($T-P$) (a) and temperature-volume ($T-V$) (b) coordinates.

Obviously, these necessarily simplified schemes do not reflect the whole variety of phase diagrams and phase transitions. For example, the melting curve of ice actually has a negative slope ($dT/dP < 0$), while in Fig. 1 the melting curve is depicted with a positive slope. Moreover, the melting curves can have a maximum, or, due to quantum effects, terminate at absolute zero temperature. The solid-state region (Fig. 1) can include numerous phase transitions associated with changes in their crystalline and electronic structure. Finally, in the case of special forms of interparticle interaction, the critical liquid-vapor point may not appear on the phase diagram of matter. In the above examples, phase transitions are accompanied by abrupt changes in the specific volume and entropy. Such transitions are called phase transitions of the first order, which usually occur at changing an aggregate state of matter and a radical transformation of the crystalline structure of solids. A first-order phase transition is determined

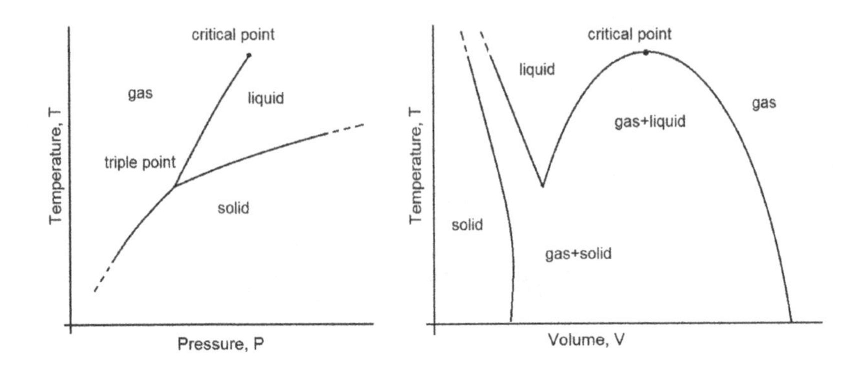

Figure 1: Typical form of phase diagram of one-component systems.

by the relations:

$$\begin{aligned} T_1 &= T_2; \\ P_1 &= P_2; \\ G_1(P,T) &= G_2(P,T), \end{aligned} \tag{1}$$

where T, P, G are the temperature, pressure, and the Gibbs thermodynamic potential, respectively. From (1) it is not difficult to derive the Clausius–Clapeyron equation, which defines the slope of the phase equilibrium curves:

$$\frac{dP}{dT} = \frac{\Delta S}{\Delta V}, \tag{2}$$

where ΔS and ΔV are the volume and entropy changes at the phase transition. Figure 2 illustrates the behavior of the Gibbs thermodynamic potential at a phase transition. As can be seen, the temperature and pressure of the phase transition are determined by the intersection of the potential branches characterizing the coexisting phases of matter. The discontinuities of entropy and volume follow from the crossing of the branches of the thermodynamic potential. Let us pay attention to the fundamental existence of metastable states of matter corresponding to potential regions, which are denoted by the dashed lines.

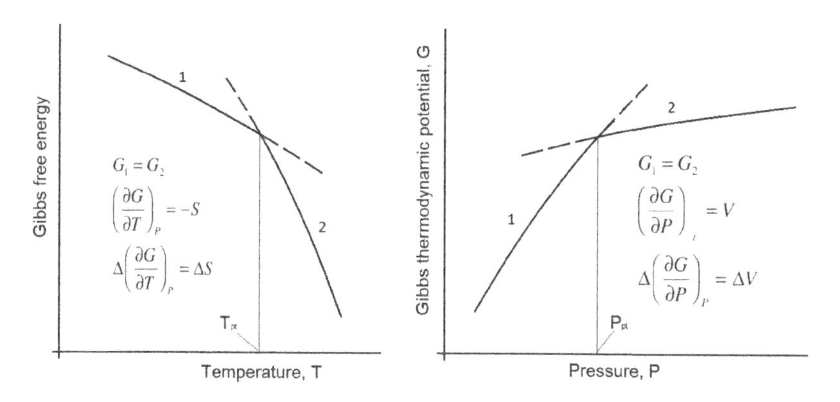

Figure 2: Behavior of the Gibbs thermodynamic potential at a first-order phase transition.

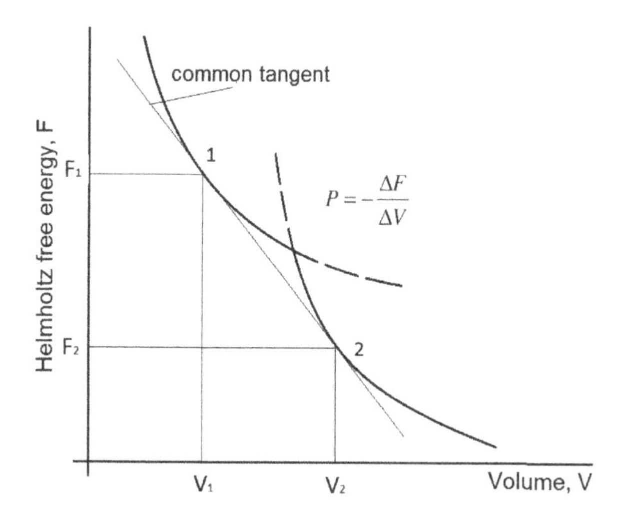

Figure 3: Behavior of the Helmholtz free energy at a first-order phase transition.

In a number of cases it is convenient to consider a phase transition using the Helmholtz free energy as a function of volume. Then, as follows from Fig. 3, the parameters of the first-order phase transition are determined by the common tangent to the branches of the free energy, and not by their intersection, as is the case for the thermodynamic potential. We recall that the general tangent condition means the equality of pressures during the phase transition.

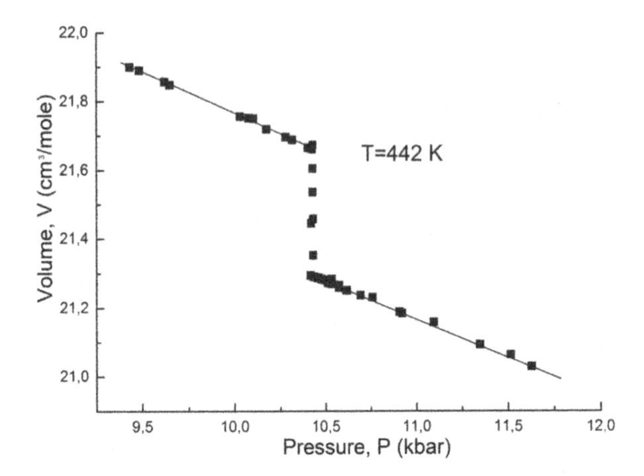

Figure 4: Volume change at the melting-crystallization of sodium, measured by means of a piston piezometer (T = 442.5 K) [1].

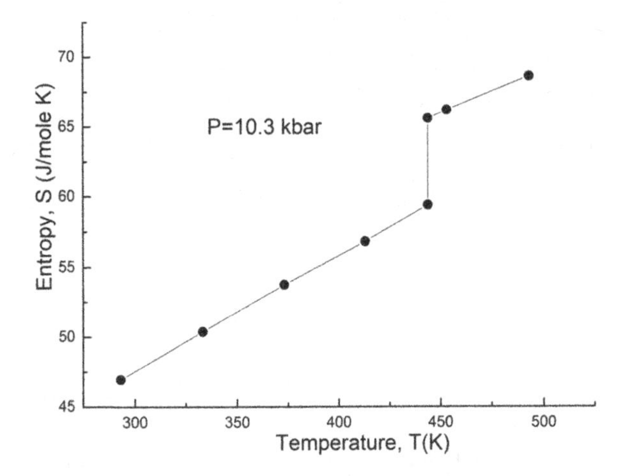

Figure 5: Entropy change at melting of sodium (10.3 kbar). The behavior of entropy is calculated on the basis of P-V-T data [2]. The entropy change is determined using Eq. (2).

The above-mentioned properties of the first-order phase transitions can be illustrated well in Figs. 4 and 5, which characterize the behavior of volume and entropy at melting of sodium, obtained many years ago in the author's laboratory.

Along with the first-order phase transitions, there is also an extensive group of phase transitions of second order or continuous phase transitions, characterized by a continuous change in the specific volume and entropy, but often accompanied by the peculiarities of heat capacity, coefficient of thermal expansion, compressibility, etc. The second-order phase transitions include transitions associated with an emergence of magnetism, superconductivity, superfluidity, orientational order, etc.

This classification, proposed in due time by P. Ehrenfest, operates with the orders of the derivatives of thermodynamic potential that experience discontinuities at the phase transition. Eliminating the uncertainty in Eq. (2) for $\Delta S = 0$ and $\Delta V = 0$ by the L'Hôpital's rule, Ehrenfest proposed equations relating the slope of the phase transition curve to discontinuities in the heat capacity, compressibility, and thermal expansion coefficient [5].

Subsequently, it was found that, for the most part, in the phase transitions of second order, the divergence of the corresponding quantities is observed as a result of fluctuation effects. Nevertheless, Ehrenfest's classification and its equations not only retain their significance as a way of systematizing phase transitions, but can also be used in analyzing phase transitions with weakly developed fluctuations, such as superconducting phase transitions, occurred in the conventional superconductors. The results of measurements of the heat capacity and thermal expansion of white tin, demonstrating completely "Ehrenfest" behavior, are displayed in Figs. 6 and 7.

It should also be pointed out that there exists in some sense an intermediate group of phase transitions — the so-called first-order phase transitions close to the second-order ones. In this case, the phase transition initially occurs according to the scenario of the continuous transitions, but somewhere, not reaching the virtual point of the second-order phase transition, ends abruptly as a first-order transition. The phase transitions in ferroelectrics $BaTiO_3$ and

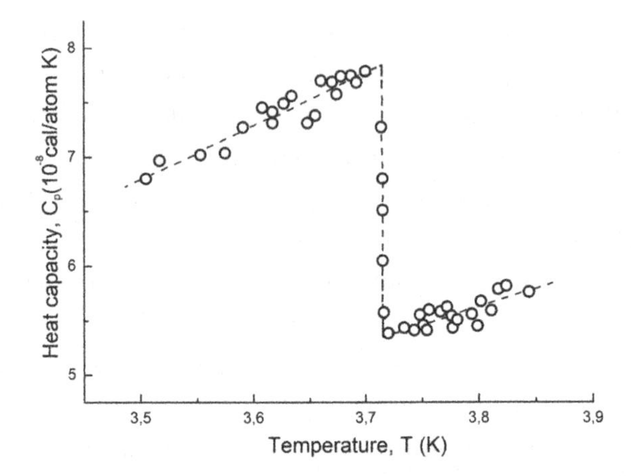

Figure 6: Heat capacity of tin in the region of superconducting phase transition [3].

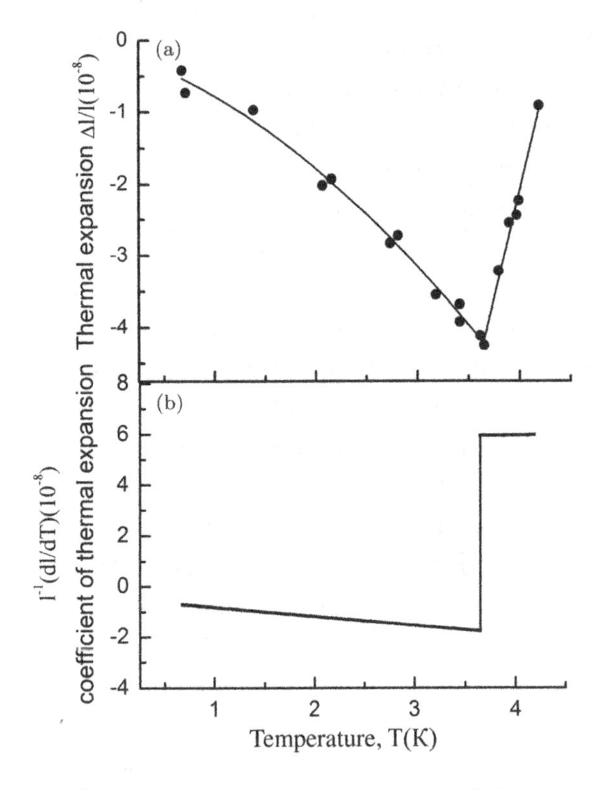

Figure 7: Linear thermal expansion of a single crystal of tin in the region of the superconducting phase transition [4].

KH_2PO_4 (KDP) can serve as examples. It should also be noted that in two-dimensional and quasi-two-dimensional systems, Kosterlitz–Thouless transitions can be observed that are not accompanied by distinct features of thermodynamic quantities. Consequently, they are dropping out of the Ehrenfest's classification.

Chapter 2

Van der Waals model and critical phenomena

2.1. Van der Waals equation

The year of 1869 should be considered as a starting point in the development of scientific ideas about phase transitions, when Thomas Andrews unveiled his studies of critical phenomena in CO_2. Four years later, van der Waals created his famous molecular theory of critical phenomena, which has not lost its significance up to the present time. In view of the particular importance of this theory, its main points are outlined.

The standard van der Waals equation has the form:

$$\left(P + \frac{a}{V^2}\right)(V - b) = RT. \tag{3}$$

It is convenient to rewrite (3) as:

$$P = \frac{RT}{V - b} - \frac{a}{V^2}. \tag{4}$$

The original idea of van der Waals was to represent the interparticle interaction in the form of an infinite repulsion at some finite distance and a very long-range attraction, which could be described simply as a rectangular well. We note that the assumption of an extremely long-range character of the attractive forces is excessive, but this allowed van der Waals to calculate, without difficulty, the contribution of the attractive interaction to the energy and pressure of the system. To take into account the contribution of the repulsive interaction, van der Waals had to solve the problem of hard spheres exactly. With this task, van der Waals, like many of his followers, failed. To be fair, we note that the exact solution of this problem is also known, at present, only for the one-dimensional case. As it turned out, the solution of the problem of solid particles, obtained by van der Waals, corresponds to the one-dimensional case, which, however, did not prevent him from describing the main features of the phenomena associated with boiling and condensation. However, taking into account the above, Eq. (4) should be rewritten in the form:

$$P = \frac{RT}{V} f\left(\frac{V}{V_o}\right) - \frac{a}{V^2}. \tag{5}$$

Going somewhat forward, we will point out that, as shown by numerical studies, the function $f(V/V_o)$ reveals an irregular behavior in the region of reduced volumes $1.5 \geq V/V_o \geq 1.36$, indicating an existence of the first-order phase transition in systems of hard spheres associated with melting-crystallization of these solid particles. We will return later to the problem of melting-crystallization, and now we turn again to the questions related to the boiling-condensation and the critical point.

Figures 8 and 9 depict experimental and van der Waals isotherms illustrating an evolution of the first-order gas-liquid phase transition and the emergence of a critical point.

As can be seen from Fig. 9, the volume change, corresponding to the phase transition, decreases with increasing temperature and disappears at the critical point. The critical temperature corresponds

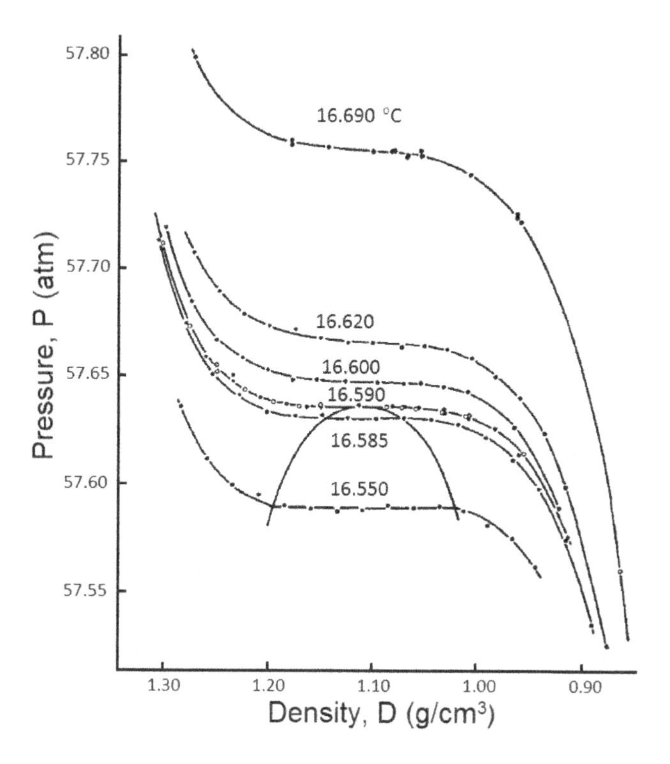

Figure 8: Compression isotherms of xenon Xe in the critical region. [6]

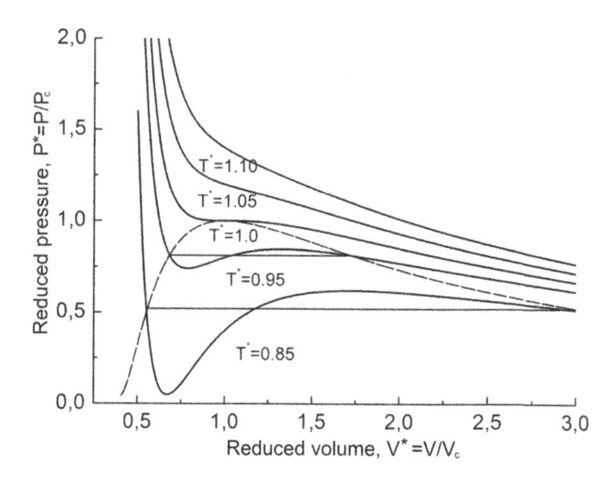

Figure 9: Van der Waals isotherms. The dashed line is the coexistence curve.

to the inflection point on the critical isotherm, where

$$\left(\frac{\partial P}{\partial V}\right)_T = 0; \quad \left(\frac{\partial^2 P}{\partial V^2}\right)_T = 0. \tag{6}$$

At temperatures above the critical, the difference between liquid and vapor disappears. It will be convenient to turn to the reduced van der Waals equation. To do this, we express the critical temperature, the critical volume, and the critical pressure through the parameters of the van der Waals equation. From (3) and (6) we obtain:

$$T_c = \frac{8a}{27b}; \quad V_c = 3b; \quad P_c = \frac{a}{27b^2}. \tag{7}$$

Further, we introduce the reduced temperature, volume and pressure respectively as: $T^* = T/T_c; V^* = V/V_c; P^* = P/P_c$.

In the reduced units, the van der Waals equation becomes:

$$P^* = \frac{8T^*}{3V^* - 1} - \frac{3}{V^{*2}}. \tag{8}$$

We emphasize that Eq. (8) is the essence of the law of the corresponding states.

Differentiating (8) in volume, we have:

$$\frac{\partial P}{\partial V}|_{V=V_c}\frac{V_c}{P_c} = -6\left(\frac{T}{T_c} - 1\right). \tag{9}$$

It follows from (9) that the compressibility of $k_T = -(1/V)(\partial V/\partial P)_T$ diverges as $T \to T_c$ by the law $(T - T_c)^{-1}$. It is also easy to see that $(\partial P/\partial T)_V$ has a finite value at the critical point. This, in accordance with the thermodynamic identity

$$\left(\frac{\partial P}{\partial V}\right)_T \left(\frac{\partial T}{\partial P}\right)_V \left(\frac{\partial V}{\partial T}\right)_P = -1,$$

implies that the coefficient of thermal expansion diverges at the critical point by the same law, as the compressibility. Next, let's see how the value $\Delta V = V_{gas} - V_{liq}$ changes as one approaches the critical point along the phases coexistence curve. For this we write:

$$P^* = \frac{8T^*}{3V_{liq}^* - 1} - \frac{3}{V_{liq}^{*2}} = \frac{8T^*}{3V_{gas}^* - 1} - \frac{3}{V_{gas}^{*2}}. \tag{10}$$

Solving (10) with respect to T^*, we obtain:

$$T^* = \frac{(3V_{liq}^* - 1)(3V_{gas}^* - 1)(V_{gas}^* - V_{liq}^*)}{8V_{gas}^* V_{liq}^*}. \tag{11}$$

Further, taking into account that near the critical point $\Delta V = V_{gas} - V_{liq}$ is a small quantity ε, we introduce the notation:

$$V_{gas}^* = 1 + \frac{\varepsilon}{2}; \quad V_{liq}^* = 1 - \frac{\varepsilon}{2}. \tag{12}$$

Substituting Eq. (12) into Eq. (11), we obtain:

$$T^* = \frac{9\left(1 - \frac{\varepsilon^2}{4}\right) - 5}{4\left(1 - \frac{\varepsilon^2}{4}\right)^2}. \tag{13}$$

Expanding (13) in a series in powers of ε^2, we have:

$$T^* = 1 - \frac{1}{16}(V_{gas}^* - V_{liq}^*)^2. \tag{14}$$

We rewrite (14) in the form:

$$\frac{V_{gas} - V_{liq}}{V_c} = 4\left(1 - \frac{T}{T_c}\right)^{1/2}$$

or, simplifying, we will write down

$$V_{gas} - V_{liq} \sim (T_c - T)^{1/2}. \tag{15}$$

At the same time, since at the critical point $(\partial P/\partial V)_T = 0$; $(\partial^2 P/\partial V^2)_T = 0$, the pressure expansion in powers of $V - V_c$ gives for the pressure versus volume in the vicinity of the critical point:

$$P - P_c \sim (V - V_c)^3. \tag{16}$$

Finally, starting from the form of the internal energy of the system of van der Waals particles (see Section 2.3), $U = -(a/V) + (3/2)kT$, for the heat capacity at constant volume for $T > T_c$ we have $C_V = (3/2)k$. Accordingly, we can write

$$C_V \sim (T - T_c)^0. \tag{17}$$

On the other hand, from thermodynamics follows

$$C_P - C_V = -\frac{(\partial P/\partial T)_V^2}{(\partial P/\partial V)_T}, \tag{18}$$

allowing us to conclude, when taking into account (16), that the heat capacity C_P diverges at the critical point.

In the two-phase region for $T < T_c$, the heat capacity C_V is given by the expression (see the conclusion [7])

$$C_V - \frac{3}{2}k \simeq \frac{9}{2}k\left[1 - \frac{28}{25}\frac{(T_c - T)}{T_c}\right]. \tag{19}$$

Thus, C_V experiences a sharp change at the critical point, as illustrated in Fig. 10.

2.2. Critical exponents

At present time, when describing the critical phenomena and second order phase transitions, it is customary to represent the temperature dependence of the thermodynamic quantities in the critical region by a power function of the form $(T - T_c)^x$, where the exponents x are denoted by letters of the Greek alphabet. Accordingly, for the van

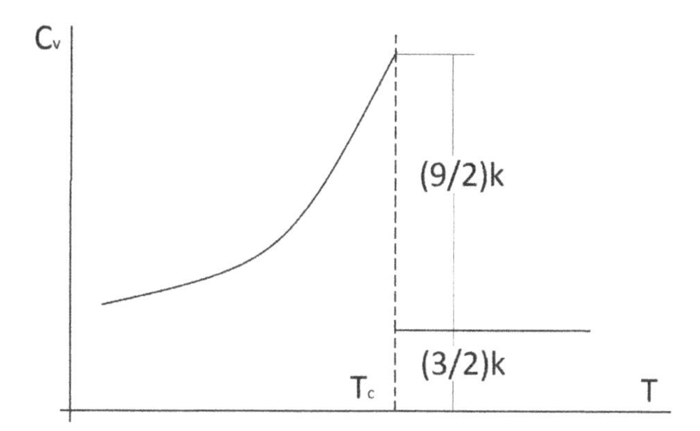

Figure 10: Change of the heat capacity at the critical point of van der Waals gas along the critical isochore.

der Waals model we can write:

$$
\begin{aligned}
C_V &\sim (T - T_c)^{\alpha}; \quad && \alpha = 0 \\
V_{gas} - V_{liq} &\sim (T_c - T)^{\beta}; \quad && \beta = 1/2 \\
k_T = -\frac{1}{V}\left(\frac{\partial V}{\partial P}\right)_T &\sim (T - T_c)^{-\gamma}; \quad && \gamma = 1 \\
P - P_c &\sim (V - V_c)^{\delta}; \quad && \delta = 3.
\end{aligned}
\tag{20}
$$

We note that it is often more convenient to operate in the relations (20) with dimensionless arguments of the form $\tau = (T - T_c)/T_c$.

A critical exponent describing the behavior of a certain function near the phase transition point is determined from the limiting relation:

$$
x \equiv \lim_{\tau \to 0} \frac{\log f(\tau)}{\log \tau}.
$$

Correspondingly, the critical exponents of "nonsingular" quantities, quantities with a logarithmic divergence or with a "peak" singularity turn out to be zero. As noted earlier, the van der Waals theory is based on the idea of the pair nature of the interparticle interaction characterized by infinite repulsion at small distances and long-range attraction. Of the real systems, noble gases most closely correspond to the indicated characteristics, so we can expect a close similarity of the phase diagrams of noble gases in the liquid-vapor equilibrium region to the van der Waals model.

Indeed, the reduced liquid-gas coexistence curve for noble gases and a number of molecular substances with the van der Waals interaction, constructed by A. Guggenheim [8], at first sight differs little from the model curve (Fig. 11). In fact, the Guggenheim curve is characterized by a very important difference from the curve following from the van der Waals model. The fact is that when the experimental curve was described by an expression of the form

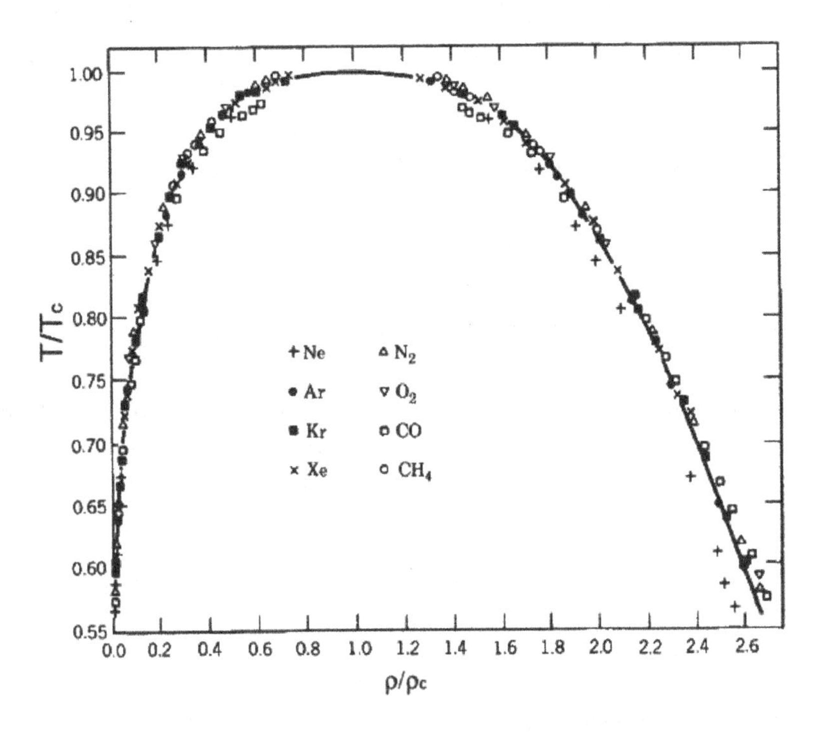

Figure 11: Reduced liquid-vapor coexistence curve after Guggenheim [8].

$V_{gas} - V_{liq} \sim (T_c - T)^{\beta}$, the exponent β turned out to be 1/3 instead of 1/2, as is the case in the van der Waals model (Eq. (20)). This circumstance played an important role in the construction of the fluctuation theory of the critical point.

2.3. Critical temperature in the van der Waals model

It is generally accepted that the critical temperature is a quantity on the order of the depth of the potential well in the interaction potential. Indeed, in the case of noble gases, the ratio $T_c/\varepsilon \simeq 1.28$ [9] favors a similar conclusion (ε is the energy parameter of the Lennard–Jones potential, which determines the magnitude of the potential well). Let's try to make an appropriate assessment for the van der Waals model. To do this, we need to express the ratio a/b in the formula $T_c = 8a/27b$ (see (7)) through the microscopic parameter that measures the depth of the potential well in the interaction potential. Let us write down the interaction potential of van der Waals particles in the form:

$$\Phi(r) = \begin{cases} \infty, & r \leq \sigma; \\ -\varepsilon(\sigma/r)^6, & r > \sigma. \end{cases} \tag{21}$$

Correspondingly, for the interaction energy, we have:

$$U = -(N^2/2V) \int_{\sigma}^{\infty} \varepsilon(\sigma/r)^6 g(r) 4\pi r^2 dr.$$

Ignoring the difference between the radial distribution function $g(r)$ and the unit for small r (this is the mean field approximation), we get:

$$U = -\frac{\varepsilon N^2}{2V}\frac{4\pi\sigma^3}{3} = -\frac{\varepsilon N v}{2(V/N)} = -\frac{a}{V_0}; \tag{22}$$

here $V_0 = V/N$. Identifying the value of Nv with the constant b and substituting the product $\varepsilon b/2$ in the expression $T_c = 8a/27b$ (7), we have:

$$\frac{T_c}{\varepsilon} \simeq 0.15. \tag{23}$$

This result shows that the naive choice of ε as a scale factor does not seem to work in the case of critical phenomena. The reasons for this situation are set out below. It is well known that the van der Waals loop marks the existence of virtual states of matter characterized by negative values of the bulk modulus. The latter means the appearance of an instability and, as a consequence, a first-order phase transition. An increase in temperature gives a positive temperature contribution to the modulus, eliminates the instability and, subsequently, a phase transition; on this way the system passes through a boundary state with a zero modulus value, called the critical point (Fig. 12).

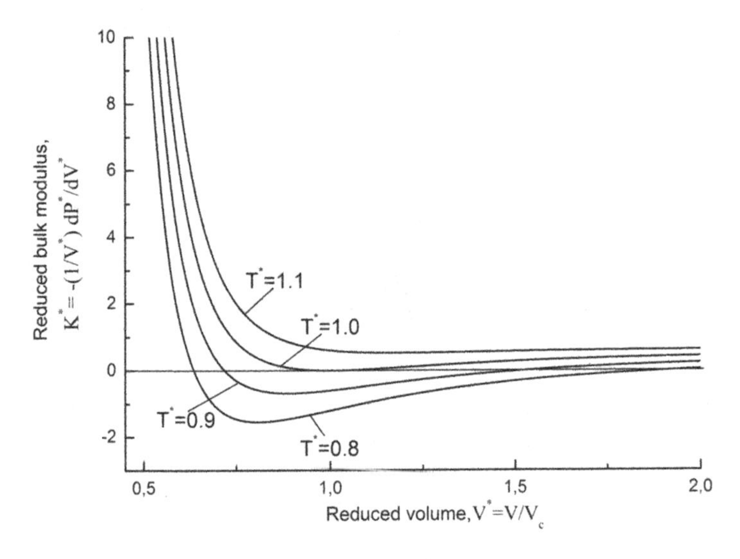

Figure 12: Evolution of the bulk modulus $K_T = -V(\partial P/\partial V)_T$ with temperature in the van der Waals model.

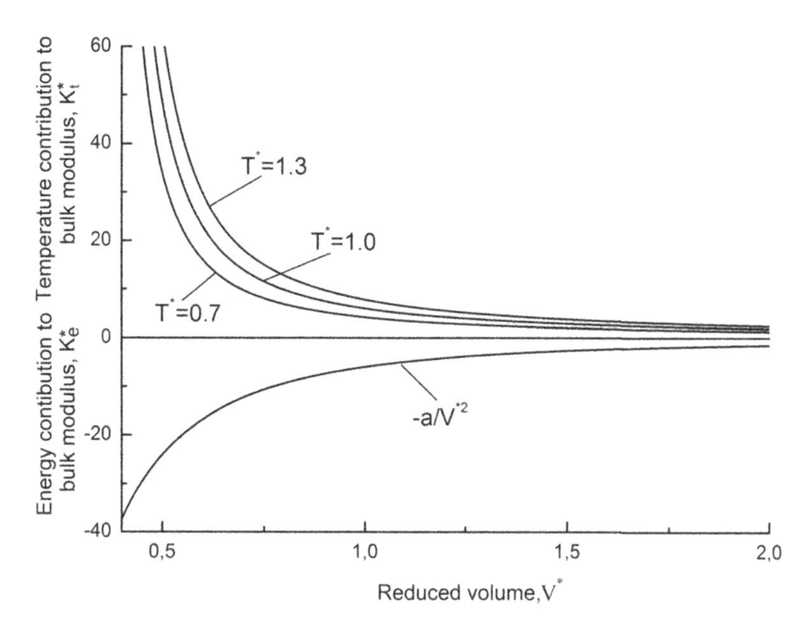

Figure 13: Temperature and energy contributions to the bulk modulus in the van der Waals model.

It is convenient to illustrate the evolution of the bulk modulus in the critical region using the van der Waals model (see Figs. 12, 13), since in the latter the thermal and potential contributions can be separated. Figures 12 and 13 present the curves describing the behavior of the modulus $K^* = (24T^*V^*/(3V^* - 1)^2) - 6/{V^*}^2$ and its constituents in the critical region in reduced units. It is easy to see that at the critical point there is an exact compensation of the thermal and potential contributions to the bulk modulus. Obviously, this situation should be very sensitive even to small variations of the interparticle interactions and the spectrum of thermal excitations. Therefore, it is hardly possible to expect similarity relations of the type $(T_c/\varepsilon) = C$, where C is a universal constant for substances with different interparticle interaction. Figure 14, which shows the coexistence curves of alkali metals, mercury and noble gases, clearly demonstrates the absence of a

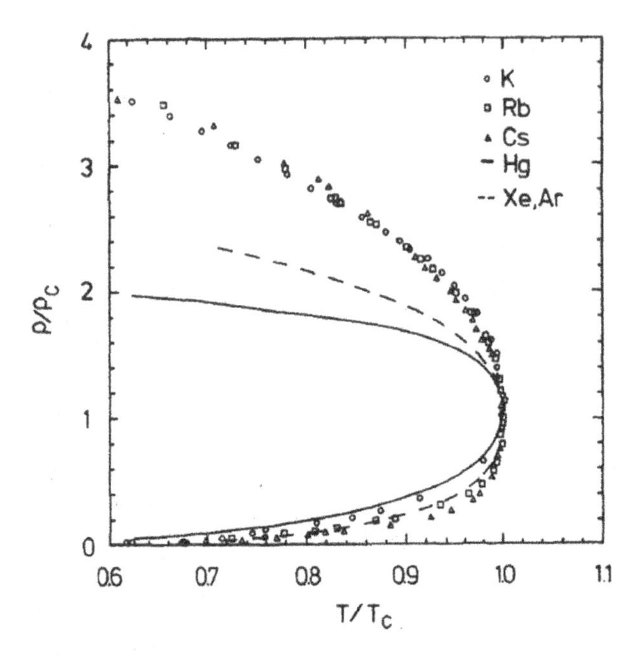

Figure 14: Liquid-vapor coexistence diagram for alkali metals, mercury and noble gases [10].

law corresponding states for different groups of substances. At the same time, as it is seen in Figs. 11 and 14, this law operates within individual groups consisting of substances with a similar character of interaction.

Recall that there is much softer interaction in the Coulomb systems than in the van der Waals model, which, however, does not interfere with the manifestation of the general macroscopic properties inherent in the phase diagram of matter presented in Fig. 1. We illustrate this situation by the example of a simple Coulomb model [13].

Let us consider the classical system of point charges on a homogeneous compensating background (the classical single-component plasma model). The total energy of such a system is written as

$$E = \frac{3}{2}kT - \frac{\alpha}{r_s},$$

where α is the Madelung constant, r_s is the radius of the sphere containing one charge, determined from the condition $(4/3)\pi r_s^3 = V/N$, where V is the volume of the system, and N is the number of charged particles. Taking into account the small variation of the Madelung constant in general, we will consider it to be independent of the thermodynamic conditions and equal to 0.9, which corresponds to the interaction of a point charge with a uniformly distributed charge of the opposite sign in a spherical cell. Note that the numerical value of α in this case does not play a special role. Further, neglecting the contribution of spatial correlations to the pressure, we obtain, for the equation of state,

$$P = \frac{0.239}{r_s^3}T - \frac{0.072}{r_s^4},$$

where r_s is expressed in Bohr radii, and the pressure P is in atomic units. The first term on the right side of the equation is nothing other than the pressure of an ideal gas, the second term is the pressure of a system of point charges. It is not difficult to verify with the above-mentioned equation that the system under consideration does not undergo a phase transition, although it has instability for small r_s. The reason for this is related to the fact that the thermal pressure $0.239T/r_s^3$ increases with decreasing r_s too slowly to overcome the instability created by the Coulomb term. However, if we assume that the compensating background is a degenerate electron gas, then the system becomes stable, and the equation of state acquires the form:

$$P = \frac{0.239}{r_s^3}T - \frac{0.072}{r_s^4} + \frac{0.1759}{r_s^5},$$

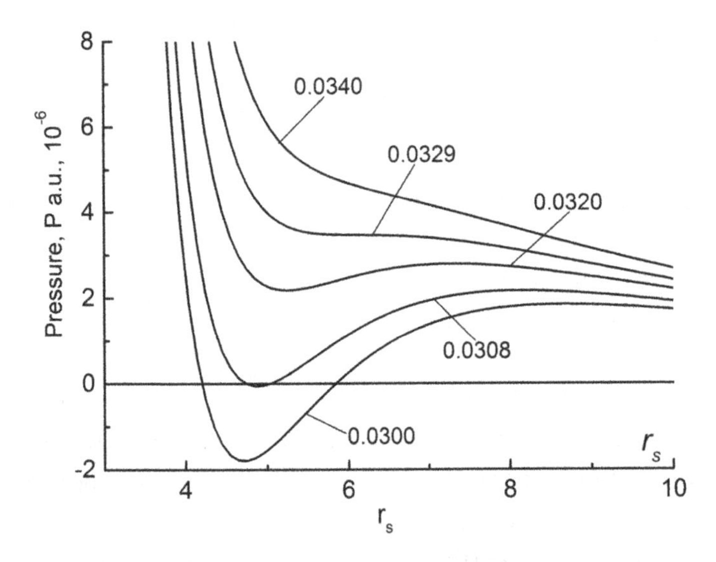

Figure 15: Van der Waals isotherms in the one-component plasma model. P is the pressure expressed in atomic units $2Ry/a_0^3$, where Ry is the Rydberg constant, a_0 the Bohr radius, and r_s is the radius of the sphere, containing one particle, expressed in units of a_0. The temperatures at the isoterms are expressed in double Rydbergs. The critical temperature corresponds to the value $T = 0.0329$ [13].

where the third term on the right-hand side is the Fermi pressure of the electron gas.

As follows from Fig. 15, the equation obtained has all the features similar to the van der Waals equation. The critical point has coordinates $T_c = 0.033$ $a.e.$ (10^4 K). The phase transition in the present model, apparently, is a liquid-liquid transition. Obviously, in the region of a first-order phase transition, these two liquid phases differ from one another only in density. Note that the repulsion in this model is volume dependent and has nothing to do with the interparticle interaction.

It is interesting that the exotic form of matter — a dense system of excitons with an interaction character close to that in alkali metals — also reveals a phase diagram of the van der Waals type (Fig. 16). Here it should be noted that Landau and Zel'dovich [11] at one time discussed the idea of the existence of two critical points corresponding to the "liquid-gas" and "metal-insulator" transitions that may arise in liquid metals at evaporation. The experimental data, however, support a single critical point, a case also considered by Landau and Zel'dovich. In concluding this section, we note that since the existence of a liquid-vapor phase transition and the critical point is due to the behavior of only the second derivative of the thermodynamic potential or free energy, it can be assumed that, for certain specific interactions, other forms of phase diagrams can also be realized [13].

Indeed, an unusual situation was observed in the simulation of the phase diagram of fullerite C60 [14]. As it turned out, there is no liquid phase on the model phase diagram of fullerite together with its indispensable attributes: triple and critical points (Fig. 17).

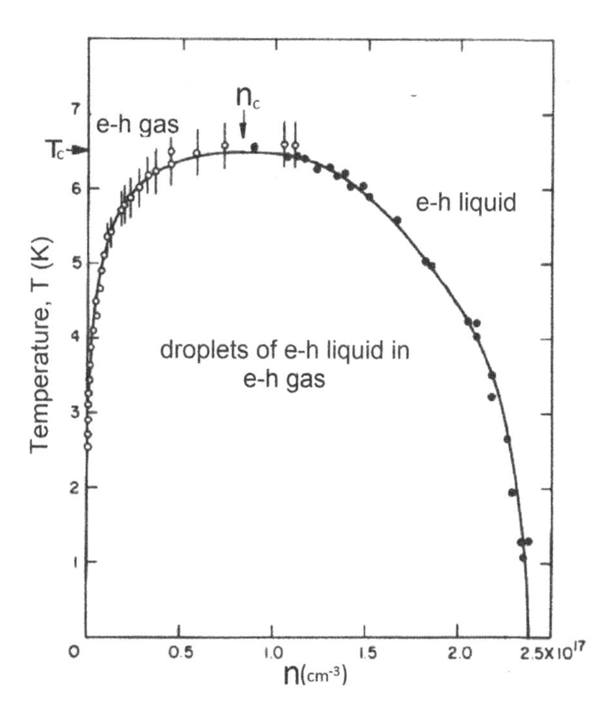

Figure 16: Liquid-gas coexistence curve for the exciton system [12].

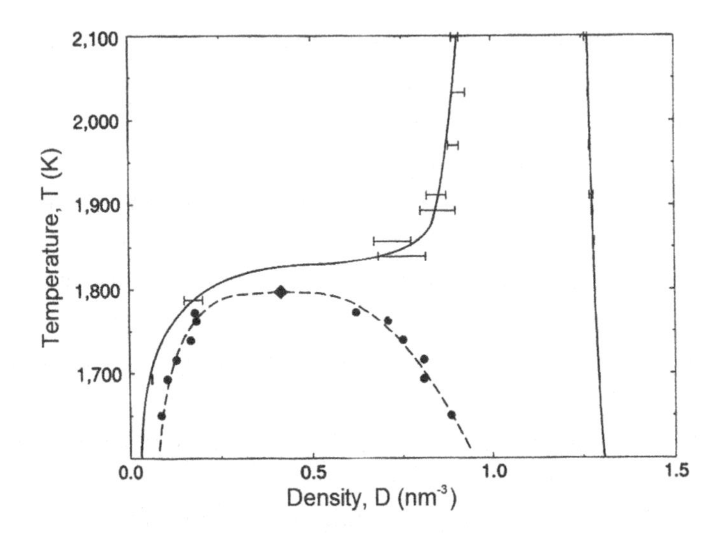

Figure 17: Phase diagram of the fullerite-C60 model [14]. Solid lines represent the curves of coexistence for the crystal-fluid equilibrium. The dashed line is the metastable liquid-vapor coexistence curve. It can be seen that the critical point is located below the crystal-liquid coexistence curve, which means that there is no liquid phase in the C60 model.

Later it was shown that by varying the parameters of the interparticle potential that determine the length of the attractive interaction, a series of phase diagrams, both containing and not containing the liquid phase, can be obtained. Curiously, in the case of a very short-range potential in the phase diagram, an isostructural phase transition arises in the stability field of the crystalline phase, terminating at the critical point [15]. The exotic phase diagrams under consideration are not found in systems of microparticles at the level of atoms or simple molecules, however, apparently, they can be realized in a medium of macromolecules and interacting macroparticles (colloids).

Above, considering the van der Waals model, we ignored the critical light scattering (critical opalescence) observed in the critical phase transitions. The key to the analysis of this phenomenon lies in the relation (9), which establishes the divergence of the compressibility at the critical point. In turn, the compressibility is related to the mean-square fluctuation of the number of particles in a given volume (see [5]), which actually determines the critical opalescence. A sequential analysis of the nature of critical opalescence was carried out for the first time by Orshtein and Zernike [16]. We will discuss this issue in Section 5.3.

Chapter 3

Melting

After the discovery of the continuity of the gaseous and liquid states of the substance and the critical point on the boiling curve, attempts were made to also detect the critical point on the melting curve. At one time F. Simon, using high-pressure techniques, investigated the melting curve of helium up to a temperature of 60 K, which is more than 10 times higher than its value at the critical point of liquid-vapor, and found no evidence in favor of the existence of a critical point on the melting curve (at present the melting curve of helium is known at least up to 600 K [17]). L. Landau noted that, in view of the difference in symmetry, any continuous transitions (such as transitions between a liquid and a gas in the supercritical region) between a liquid and a crystal are absolutely impossible [18]. L. Landau also showed in the framework of the "weak crystallization" model that the transition between a liquid and a crystal can generally occur only as first order [18]. According to Landau, a continuous or second-order phase transition at crystallization is possible only at isolated points, but such cases are still unknown.

Thus, the melting and crystallization of matter as a first-order phase transition can exist at arbitrarily high pressures and temperatures. We emphasize that the ability to crystallize and melt is a universal property of systems of many particles, regardless of their specific nature.

3.1. Melting of hard spheres

We will begin the analysis of melting with the simplest model of interacting particles — a system of hard spheres (see Fig. 18). The existence of crystallization in two-dimensional and three-dimensional systems of hard spheres is quite obvious at the intuitive level. In the light of the Landau analysis [18], this transition should be first order, although it was sometimes assumed in the literature that, in the absence of interaction energy, this transition can be second order [19].

We note that the system of solid particles is unstable for $P = 0$, since in the absence of an attractive interaction it is kept from expanding at $T > 0$ only by external pressure. In 1957, results of computer experiments, confirming the idea of existence of a phase transition in a system of hard spheres, were published in a single issue of J. Chem. Phys. [20,21]. Observations of the coordinates of the particles left no doubt that the detected phase transition is melting-crystallization. In fact, two branches of the equation of state were

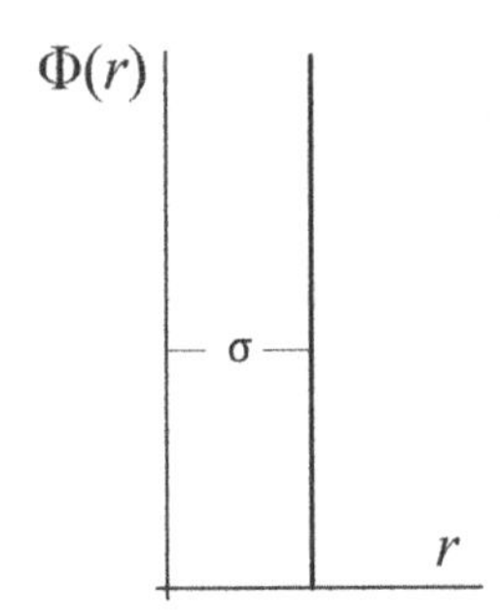

Figure 18: Interaction potential for a system of hard spheres.

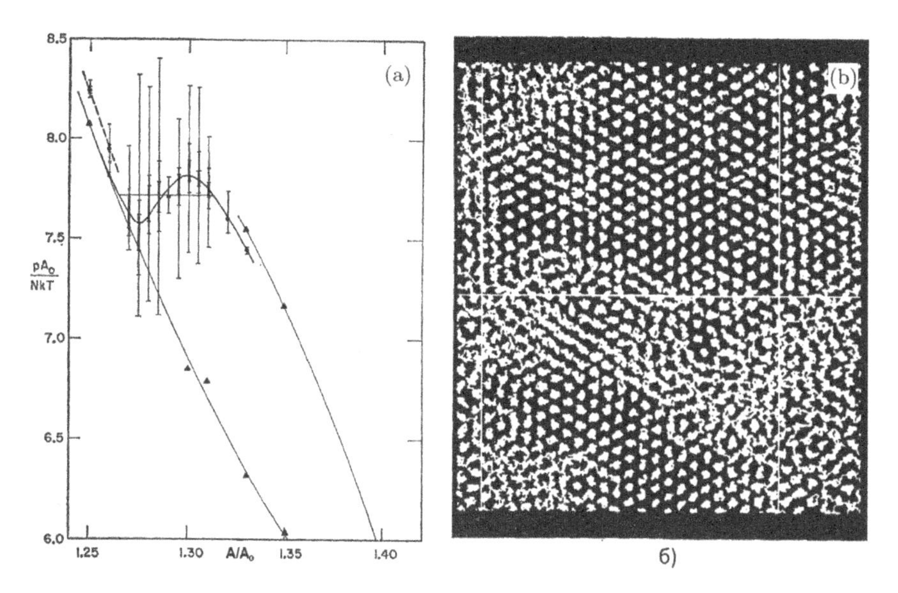

Figure 19: a) Compression isotherm of a solid disk system, demonstrating an existence of phase transition, b) Particle trajectories during a phase transition in a solid disk system [22].

observed, corresponding to the liquid and crystalline state of the system. However, the coexistence of phases was later observed in a two-dimensional system (Fig. 19). The weak logarithmic divergence of the mean-square displacements of particles does not prevent the realization of melting in a two-dimensional crystal as a first-order phase transition.

Further, following L.K. Runnels [23] (see also [24]) we show that just a fact of existence of a phase transition in a system of solid particles is sufficient to obtain important information about its properties. Let us write the free energy of the classical system of solid particles in the form

$$F = U - TS = \frac{3}{2}kT - TS, \tag{24}$$

hence, for pressure follows

$$P = T\left(\frac{\partial S}{\partial V}\right)_T. \tag{25}$$

Further, using the Maxwell relation $(\partial S/\partial V)_T = (\partial P/\partial T)_V$ one gets

$$(\partial P/\partial T)_V = P/T$$

or

$$dP/P = dT/T,\ V = const.$$

Integrating, we conclude that $lnP = lnT + C$, or, since the constant C can depend only on the specific volume or density of the system, we finally obtain

$$\frac{P}{kT} = \Gamma(\rho), \tag{26}$$

where $\Gamma(\rho)$ — some function of density.

Suppose that in a system of solid particles a first-order phase transition occurs (which actually does). Then, from the condition of equality of Gibbs potentials at a phase transition, we have $P\Delta V = T\Delta S$, whence, using the Clausius–Clapeyron equation $dP/dT = \Delta S/\Delta V$, we obtain the phase transition equation

$$dP/dT = P/T$$

and

$$P = cT. \tag{27}$$

The constant c in this equation has the dimension of density. In this connection, it is reasonable to rewrite this equation in the form:

$$P = \frac{c'}{\sigma^3}T, \tag{28}$$

where σ is the diameter of a solid sphere. The form of the resulting equation means that in the solid particle system, the phase equilibrium curves are straight lines originating from the origin.

Comparison of (26) and (27) shows that at phase transitions in the solid particle systems, the volumes of the coexisting phases V_l and V_s, the volume ΔV and the entropy discontinuities ΔS are constants. Indeed, as follows from (27), the ratio P/T is a constant along the melting curve of hard spheres, which, in accordance with (26), means the invariance of the density or volume of the system along the melting line. Consequently, the volume jump at melting $\Delta V = const$ and, since the slope of the melting curve is also the constant $dP/dT = const$, the entropy of melting ΔS of the system of hard spheres remains unchanged.

Note that if solid particles are in the form of ellipsoids, rods, disks, parallepipids, etc., one can expect a whole series of phase transitions in accordance with the hierarchy of characteristic lengths, which apparently occurs in liquid-crystal systems.

3.2. Melting of soft spheres and real simple systems

In real systems, the interaction potentials always have some softness, and therefore there are no temperature-independent characteristic lengths. We take this circumstance into account, using the simple technique described in [25]. We determine the corresponding temperature dependence from the condition:

$$\Phi(r) = kT,$$

where $\Phi(r)$ is the interaction potential. Assuming that $\Phi(r) \sim 1/r^n$, we obtain $r \sim T^{-1/n}$. Taking into account the last relation, we rewrite expression (28) in the form:

$$P = c'' T^{1+3/n}. \tag{29}$$

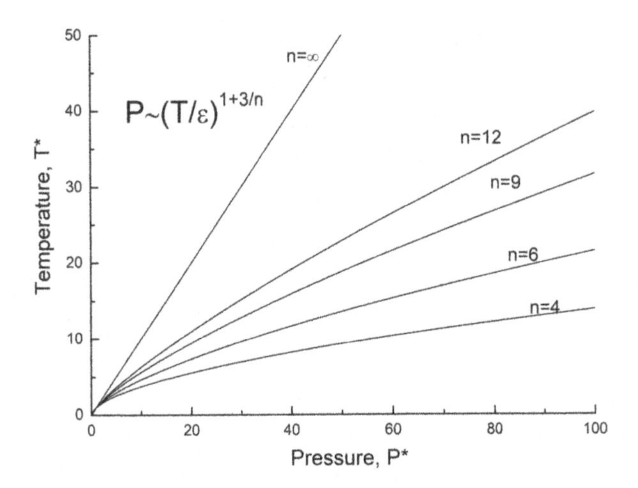

Figure 20: Melting curves of systems of particles interacting with the inverse power potential $\Phi(r) \sim 1/r^n$.

Thus, in realistic systems, the lines of phase equilibria have some curvature, as shown in Fig. 20, where the melting curves of systems of particles with repulsive interaction are presented. In real systems, there is always an attractive interaction that gives a negative contribution to the pressure, which, as we see below, leads to a simple pressure renormalization in the first approximation (see experimental melting curves in Fig. 21).

Below we consider the problem of melting, using the self-similarity of a system of particles interacting in accordance with the power law [24, 27]. Let us write down the potential energy of the system in the form:

$$U = \sum_{i<j} \Phi(r_{ij}), \tag{30}$$

where

$$\Phi(r) = \varepsilon \left(\frac{\sigma}{r}\right)^n. \tag{31}$$

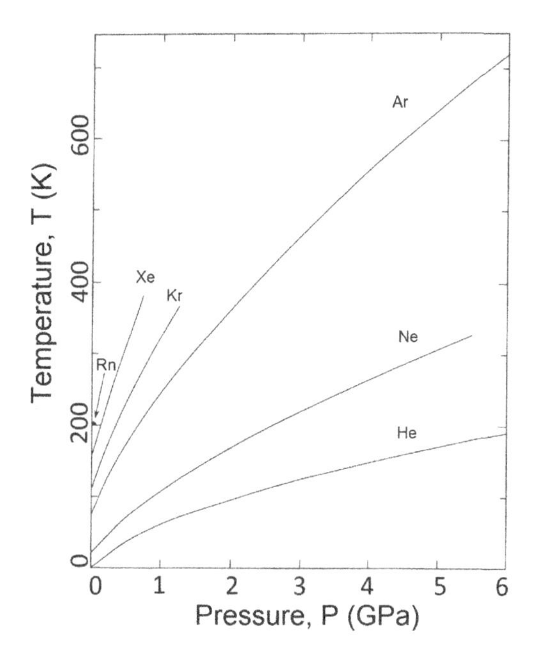

Figure 21: Melting curves of noble gases [26].

Here ε and σ are constants with dimensions of energy and length, and r is a distance between the particles. We introduce the dimensionless density $\rho = N\sigma^3$ and the dimensionless length $s = r(N/V)^{1/3}$, where V is the volume of the system. The partition function will be written as

$$Z = \frac{V^N}{N!\lambda^{3N}} \int_{s_1} \cdots \int_{s_N} exp\left(\frac{-\rho^{n/3}\sum_{i<j} s_{i,j}^{-n}}{kT}\right) ds_1...ds_N, \qquad (32)$$

where $\lambda/(2\pi mkT)^{1/2}$.

From (32) follows the equation of state

$$\frac{PV}{RT} = 1 + \varphi[\rho(\varepsilon/kT)^{3/n}]. \qquad (33)$$

We emphasize that the ratio PV/RT does not depend on temperature and volume separately, but is a function of the combined variable $\rho(\varepsilon/kT)^{3/n}$. Melting as a first-order phase transition is characterized by two values of this variable. From this follows the relations for the density of the liquid and solid phases

$$\rho_l = c_l(kT/\varepsilon)^{3/n}; \ \rho_s = c_s(kT/\varepsilon)^{3/n}, \tag{34}$$

where c_l and c_s are constants. Using (33) and (34), we obtain relations for an equation of the melting curve and changes in volume and entropy:

$$\begin{aligned} P_m &\sim \left(\frac{kT}{\varepsilon}\right)^{1+3/n}; \\ \frac{\Delta V}{V_s} &= const; \\ \frac{\Delta S}{R} &= const. \end{aligned} \tag{35}$$

We note that at $n \to \infty$, relations (34) and (35) turn into the relations characterizing the hard spheres melting. Pay attention to the fact that the relations (35) correspond to the conclusions made on the basis of simple considerations in the section on hard sphere melting. However, it should be remembered that the interaction potential in a realistic particle system always contains an attractive part (see Fig. 22). To take this circumstance into account, we use the thermodynamic perturbation theory [5]. To this end, we write the free energy F as the sum (36), where the term F_o describes the properties of a system of particles with a purely repulsive interaction, and the term $\overline{E}$ is the average value of the

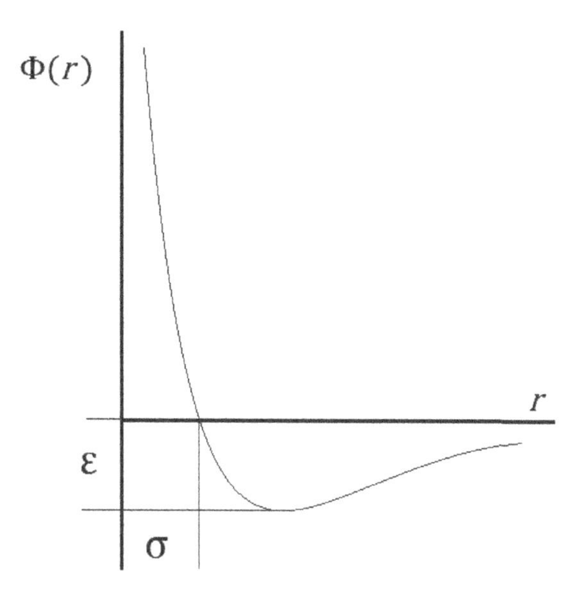

Figure 22: Realistic interparticle interaction potential.

perturbing energy, expressed with the radial distribution function $g_o(r)$ of the unperturbed system and the perturbation potential $\Phi'(r)$ (37). If the potential $\Phi'(r)$ is long-range, we can neglect the difference between the function $g_o(r)$ and unity at small r and, as a result, we obtain a correction to the free energy of the system proportional to its density (van der Waals approximation).

$$F = F_o + \overline{E}; \tag{36}$$

$$F = F_o + \frac{N^2}{2V} \int \Phi'(r) g_o(r) 4\pi r^2 dr; \tag{37}$$

$$F = F_o - \frac{a}{V} = F_o - a\rho. \tag{38}$$

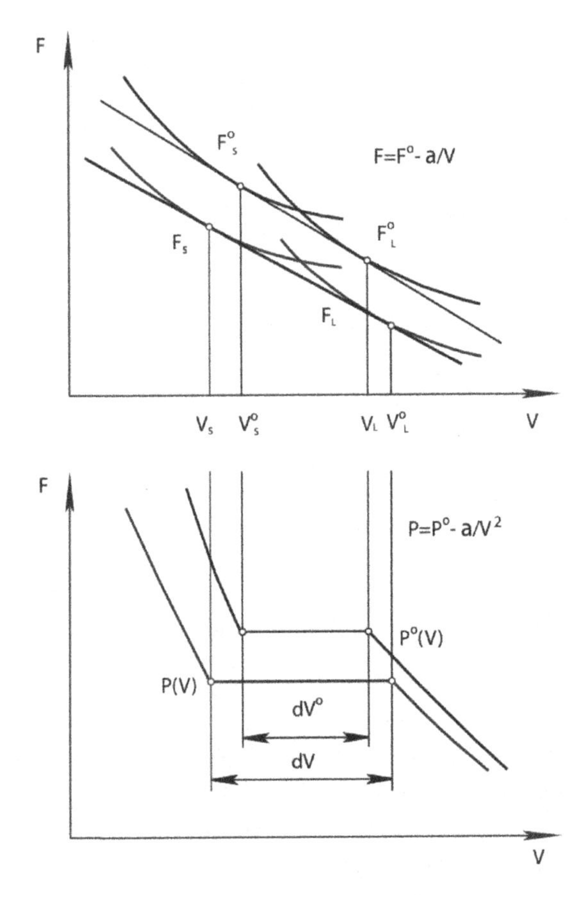

Figure 23: Influence of the attractive interaction on the pressure and volume discontinuities at melting [24].

Figure 23 shows that the attraction lowers the pressure and increases the volume change upon melting, wherein the entropy of the transition ΔS naturally grows. Nevertheless, as shown in Fig. 24, the relations (35) remain true in the high temperature limit and acquire the meaning of asymptotic values (on compression, the term F_0 grows much faster than a/V).

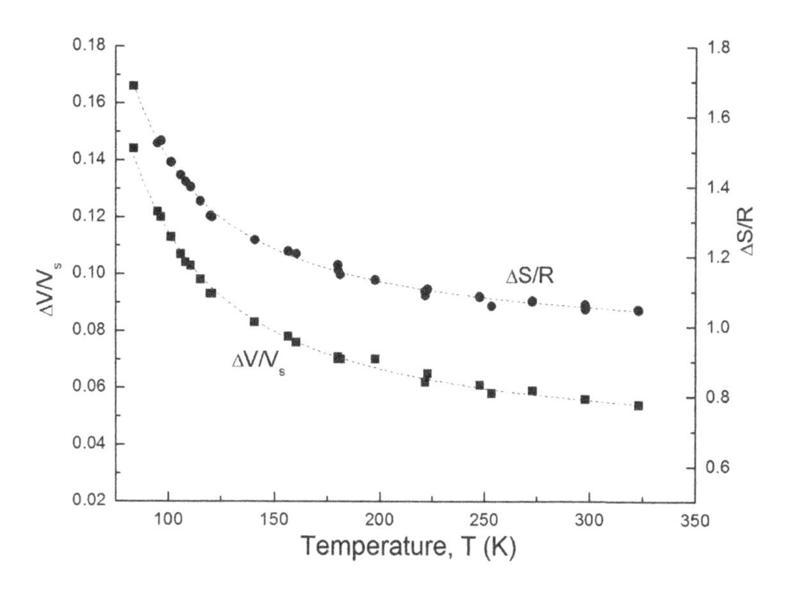

Figure 24: Entropy of melting $\Delta S/R$ and relative volume change $\Delta V/V_S$ at the argon melting [24].

3.3. Melting entropy

Figure 25 depicts the dependence between the entropy of the melting $\Delta S/R$ and the relative change volume $\Delta V/V_S$ on melting for argon, sodium and a number of systems with a purely repulsive power law interaction. As can be seen, the melting entropy tends to a constant in the limit $\Delta V/V_S \rightarrow 0$. The numerical value of this constant is close to 0.7 R, which suggests the connection of this quantity with the so-called "collective" entropy introduced in the paper [28].

The notion of "collective" entropy is illustrated in Fig. 26 (a) and (b), where the system N of noninteracting particles confined in the volume V is displayed. However, in case (a) the motion of each particle is limited to V/N, while in case (b) the full volume V is available for the motion of the particle. Finally, writing down

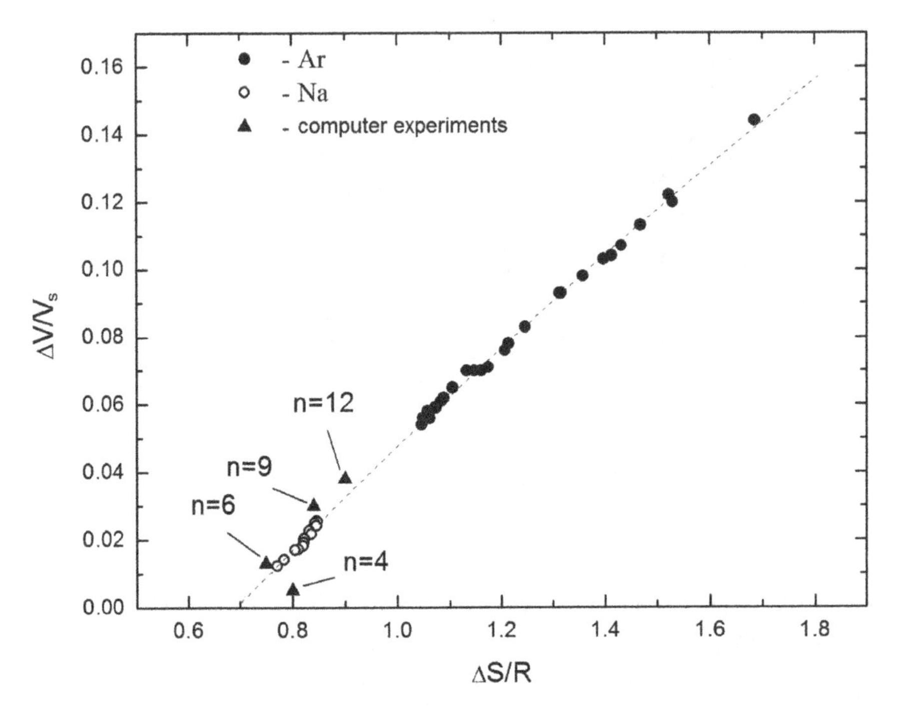

Figure 25: Interrelation between entropy of melting $\Delta S/R$ and relative change of volume $\Delta V/V_S$ at the melting of argon, sodium and a number of systems with purely repulsive power law interactions [24].

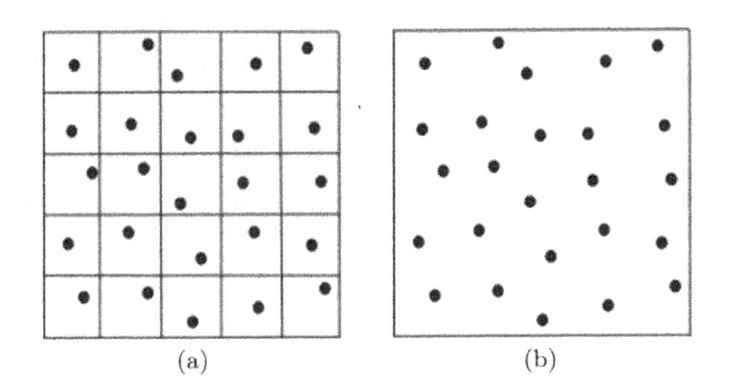

Figure 26: Illustration of the concept of collective entropy.

the partition function Z for the system in states (a) and (b) in the form (39) and taking into account the relation $S = -(\partial F/\partial T)_V = -\partial(-kT\, lnZ)/\partial T$, we obtain for the "collective" entropy the value

$$Z_1 = \frac{\lambda^{3N} V^N}{N^N}; \; Z_2 = \frac{\lambda^{3N} V^N}{N!}; \tag{39}$$

$$\Delta S = S_1 - S_2 = k \ln \frac{N^N}{N!} = Nk = R. \tag{40}$$

Chapter 4

Step potential, quantum effects and "cold" melting

4.1. "Collapsing" systems

By "collapsing" systems here are meant systems described with the help of a step potential (see Fig. 27). Systems of this type are studied in connection with anomalous melting curves, isomorphic phase transitions, transformations in colloid systems, etc. As can be seen from Fig. 27, the presence of a step immediately leads to the appearance of the energy scale ε in the system.

Let us now turn to the analysis of the "collapsing" system (Fig. 28). Obviously, in two limiting cases $T \ll \epsilon$ and $T \gg \epsilon$ the system behaves as a simple system of hard spheres with diameter σ' or σ. The transition between the two branches of the melting curve occurs at $T \approx \varepsilon$, while, depending on the value of the ratio σ'/σ, one should expect a more or less significant anomaly on the melting curve, which could show a temperature maximum or a triple point, associated with a phase transition in a liquid [29].

Next, we point out that the form of the interaction potential in the "collapsing" system suggests the possibility of an isostructural phase transition in the crystalline phase, which was observed in computer

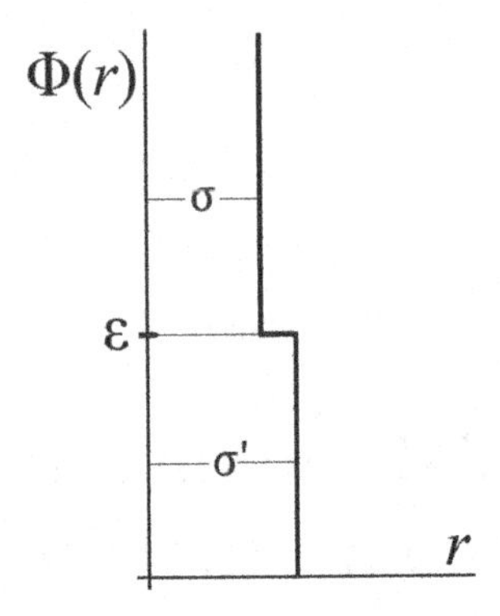

Figure 27: Step interaction potential.

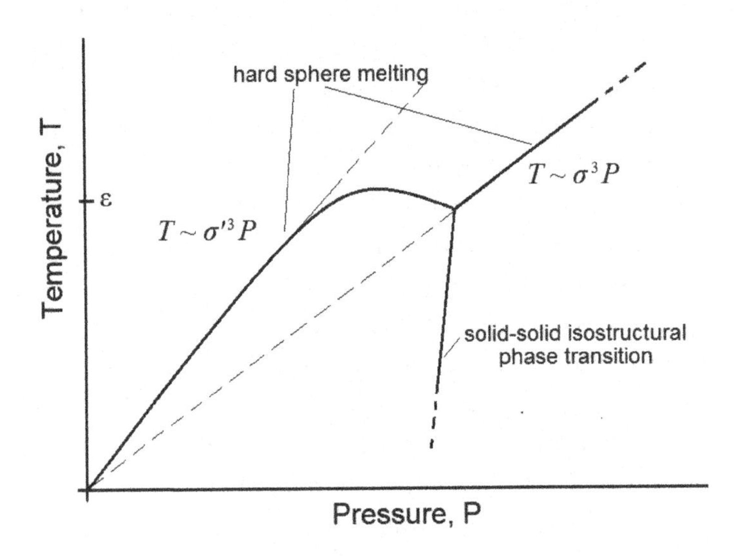

Figure 28: Schematic view of the phase diagram of a system of particles interacting via step potential.

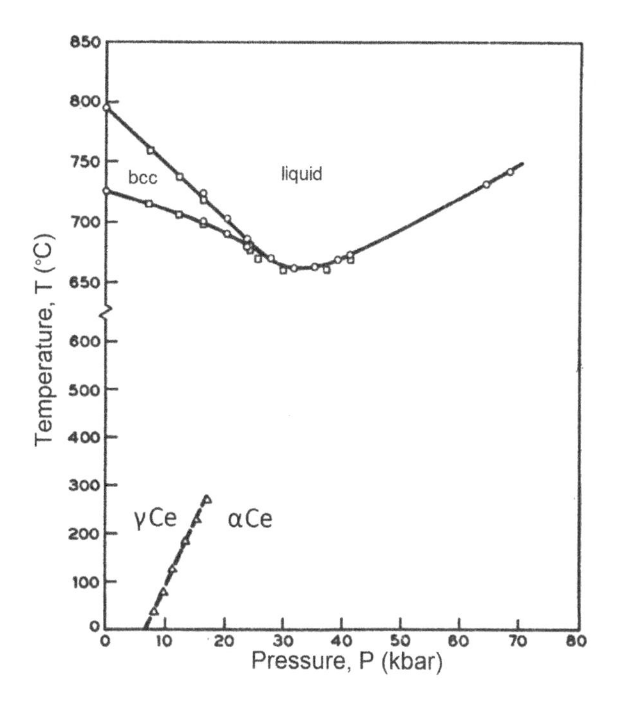

Figure 29: The phase diagram of cerium Ce according to the data of [32].

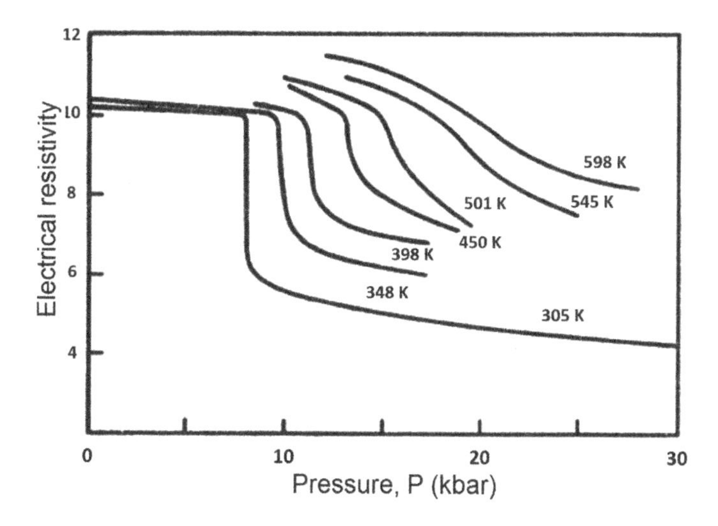

Figure 30: Electrical resistivity of Ce at high pressures and temperatures [32].

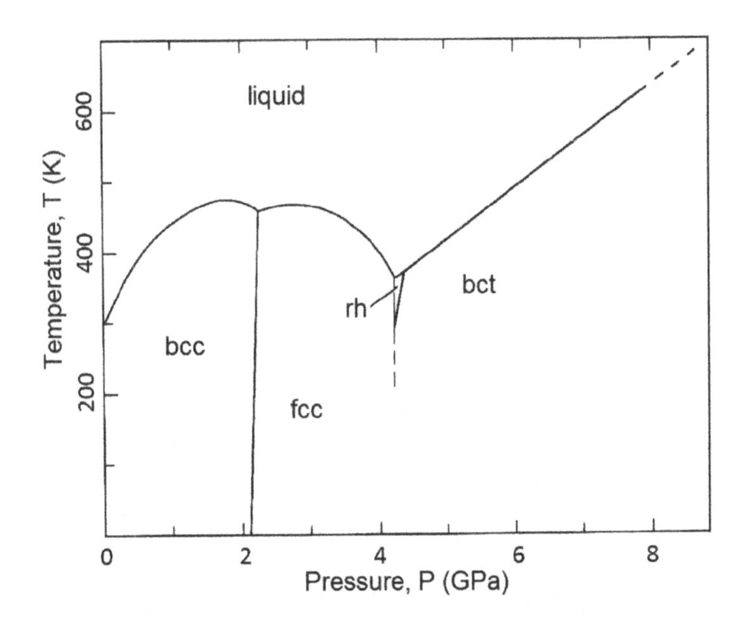

Figure 31: Phase diagram of cesium [26].

experiments [30, 31]. The specific features of the phase diagrams of cerium and cesium (maxima and minima on the melting curves, isostructural phase transitions) presented in Figs. 29 and 31 can be described in terms of two atomic states, and is modeled by a step potential. Let us pay attention to the character of the dependence of the electrical resistance of cerium on pressure (Fig. 30), which obviously points to the critical behavior of the isostructural $\gamma-\alpha$ phase transition at high temperatures. We also emphasize that the behavior of the entropy of melting during the melting of cesium (Fig. 32) can be interpreted as a manifestation of entropy of mixing arising as a result of the formation of excited states of the system.

The physical causes of the anomalies observed on the melting curves (maxima and minima) are associated with delocalized or "blurred" transformations occurring in a liquid or solid. These transformations can be associated with a change in the atomic (molecular) and/or electronic structure of liquid and solid. However,

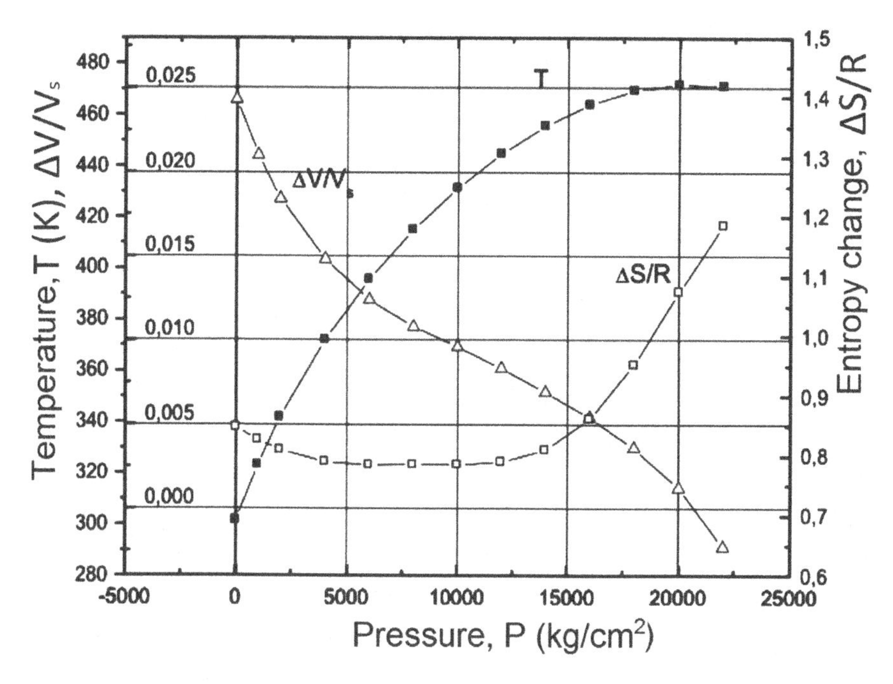

Figure 32: Behavior of changes of entropy $\Delta S/R$ and relative volume $\Delta V/V_S$ at melting of cesium Cs [33].

in the case of solid, these transformations, as a rule, manifest themselves as a first-order phase transition. The case of cerium is a kind of exception (see Fig. 29). In any case, the anomalous melting curves are a reflection of the corresponding volume anomalies leading to the equality of volumes at the extremal points. It is easy to see that the condition $\Delta V = 0, \Delta S \neq 0$ and the Clausius–Clapeyron equation $(dT/dP = \Delta V/\Delta S)$ imply $dT/dP = 0$, which corresponds to the extremal points on the melting curve (see for more details [24, 34, 35]).

4.2. Quantum effects and cold melting

There are cases when $\Delta V \neq 0, \Delta S = 0$, and $dP/dT = 0$, which occur at melting helium isotopes at low temperatures (see [36]).

Figure 33 demonstrates a pronounced pressure minimum on the melting curve ^{3}He. This behavior of the melting curve ^{3}He was predicted by Pomeranchuk as the exchange effects caused by nuclear spins; in the result of which the entropy associated with the ordering of nuclear spins decreases in the liquid with decreasing temperature faster than in the solid [37].

To conclude this section, we point out Figs. 34 and 35, illustrating the quantum effects on entropy of melting of helium isotopes. A comparison of Figs. 34 and 35 shows that the decrease in the entropy of melting of helium is closely related to an increase of the thermal de Broglie wavelength with decreasing temperature. The latter, within the framework of the quasi-classical approach, apparently indicates a reduction in the number of distinguishable configuration states of the system [35].

To speak a little further about the "cold" melting, as follows from the quantum theory, the energy of the ground state of any bound or condensed system of particles contains a dynamic part, often called "zero point" energy. The existence of the "zero point" energy is due to the Heisenberg uncertainty principle

$$\Delta x \cdot \Delta p \geq \hbar, \tag{41}$$

which establishes the connection between the uncertainty of the coordinate Δx and the momentum Δp. It follows from (41) that the localization of a particle is inevitably accompanied by an increase in its kinetic energy. For this reason, the "zero point" energy necessarily increases when the substance is compressed, but its effect on the properties of the substance also depends on the other components of the total energy. In the general case, the "zero point" energy is a decreasing function of the mass of the particles that make up the system, but its exact functional form depends on the nature of interparticle interactions and the phase state of the substance. It is obvious that "zero point" energy plays a particularly important role in the case of systems with a small mass of particles and a

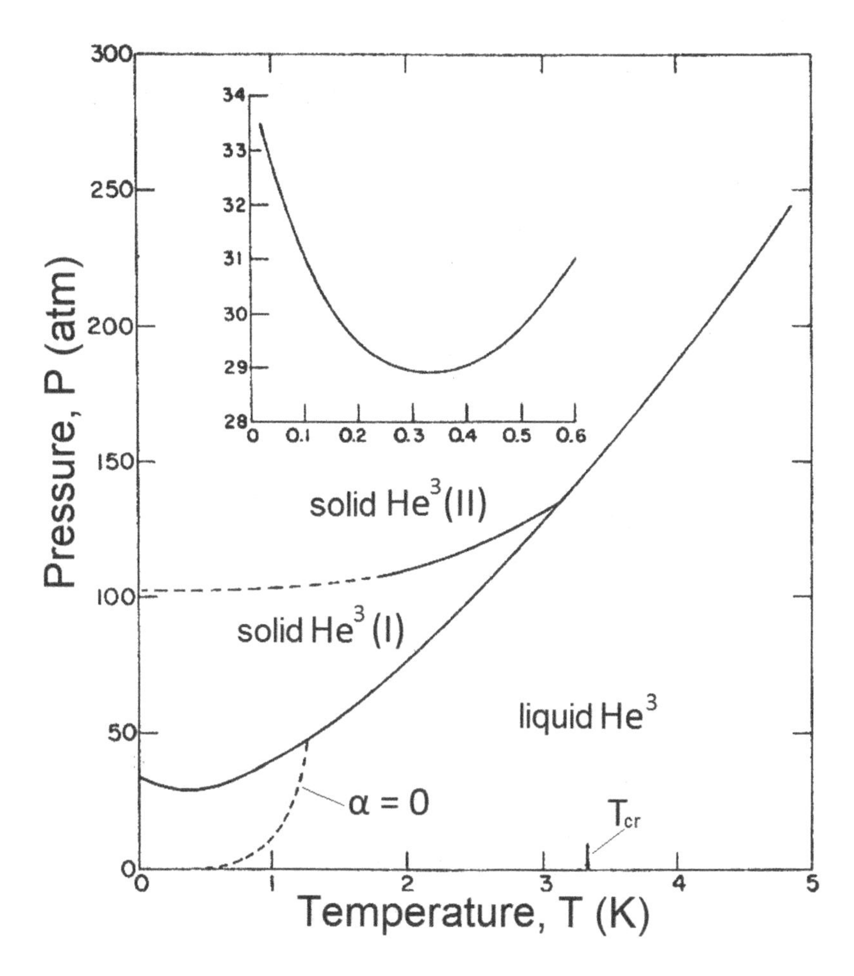

Figure 33: Phase diagram of ^{3}He [36].

relatively weak interparticle interaction. That is why helium isotopes ^{3}He and ^{4}He do not crystallize up to the lowest temperatures at low pressures as opposed to heavier noble gases. "Zero point" energy also makes a significant contribution to the equations of state of helium, hydrogen and other light substances and, obviously, has a serious influence on the processes occurring in the interior of "cold" stars and massive planets. One of the most interesting

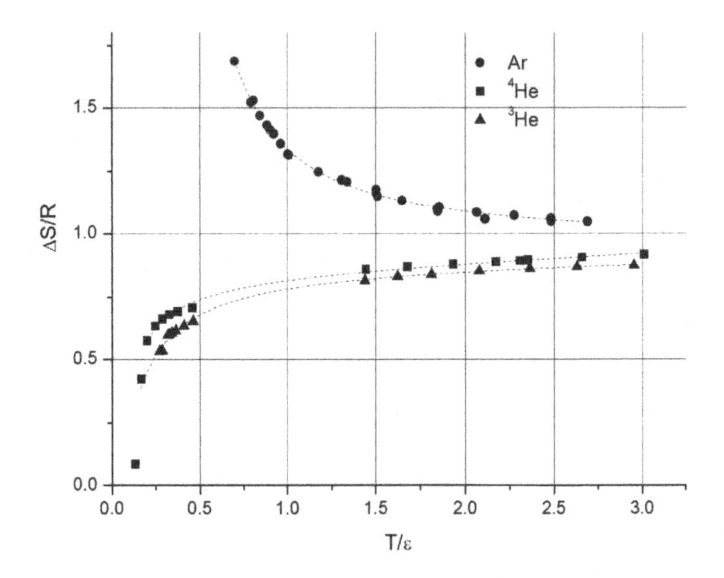

Figure 34: Entropy of melting of argon and helium isotopes as a function of reduced temperature ($\varepsilon_{Ar} = 119.3K, \varepsilon_{He} = 10.22K$) [35].

effects associated with the existence of "zero point" energy is the so-called "cold" melting of a highly compressed substance, first considered by D.A. Kirzhnits [38, 39]. Somewhat later this problem was considered by AA. Abrikosov [40]. Subsequently, the problem of cold melting was investigated in a number of papers (see the bibliography in [41, 42]). Recall that in the absence of the theory of melting, the corresponding analysis in papers [38, 40] and in many subsequent ones was actually based on the use of the Lindemann criterion [43, 44].

Previously I. Pomeranchuk [37] acted in a similar way, trying to explain the instability of the crystalline helium phase at atmospheric pressure. We recall that according to Lindemann, the crystal melts when the ratio of the rms (root mean square) amplitude of the vibrations of the atoms $\langle \delta r^2 \rangle^{1/2}$ to the interatomic distance r reaches a certain critical value, or

$$\frac{\langle \delta r^2 \rangle^{1/2}}{r} \cong C. \tag{42}$$

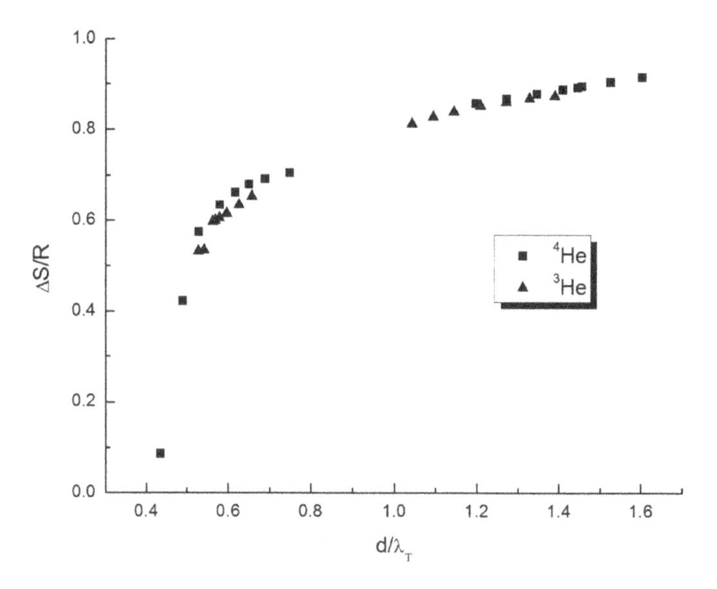

Figure 35: Entropy of melting of helium isotopes as a function of ratio d/λ_T, where d is interparticle distance, $\lambda_T = h/(2\pi mkT)^{1/2}$ is de Broglie thermal wavelength [35].

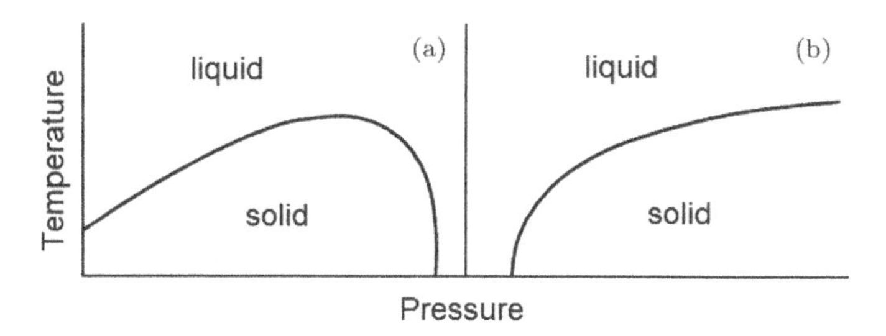

Figure 36: Influence of quantum effects on melting curves of systems with: (a) Coulomb interaction, and (b) short-range repulsion.

It is not difficult to show that in the case of a system of particles interacting according to the law $\Phi(r) \propto 1/r^n$, for zero-point oscillations, we have [37–40]:

$$\frac{\langle \delta r_s^2 \rangle^{1/2}}{r_s} \propto r_s^{(n/2)-1}; \tag{43}$$

here r_s is determined from the condition $V/N = 4/3\pi r_s^3$.

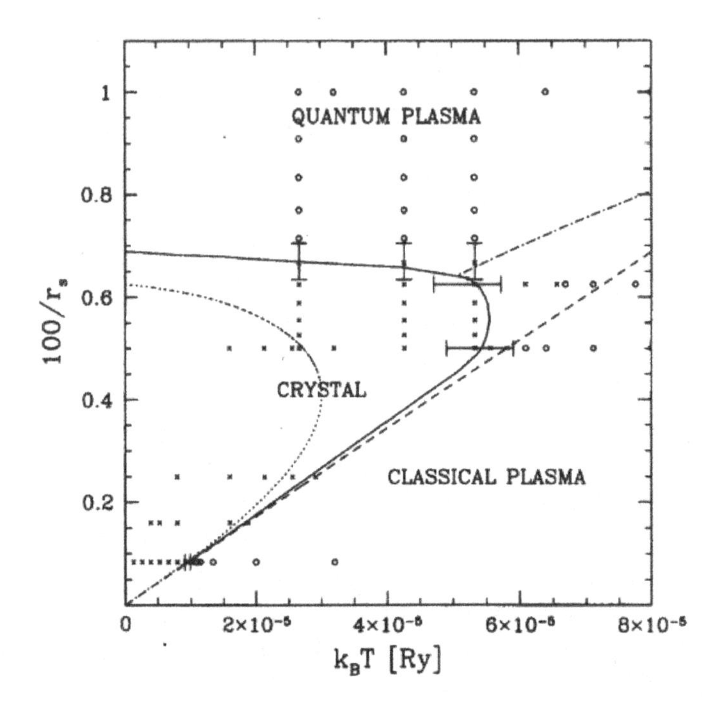

Figure 37: Phase diagram of one-component quantum plasma [42].

It follows from (43) that the relative amplitude of the "zero point" oscillations for n > 2 increases with increasing volume. In contrast, in the Coulomb case, or more precisely in the case of a quantum system of point charges on a compensating background, the relative amplitude of the "zero point" oscillations [38, 39] increases with decreasing volume. Thus, according to the ideas developed in [37, 38, 40], quantum effects lead to the emergence of two types of global phase diagram of matter, depicted in Fig. 36. We emphasize, however, that the Lindemann relation (2) is not a melting criterion, but is essentially a similarity relation valid only in the case of systems of classical particles interacting in a power law. Nevertheless, the melting of helium completely corresponds to the phase diagram shown in Fig. 36 (b). The melting of a quantum one-component plasma calculated using the quantum Monte Carlo method, shown in Fig. 37, also corresponds to the diagram depicted in Fig. 36 (a). Since in both cases the ratio of zero point and static energies is proportional to the relative amplitude of the zero-point oscillations, (43) means that the role of the "zero point" energy increases with compression in the Coulomb system and weakens in a system with a short-range power-law interaction [44]. However, the interaction in real substances not always can be reduced to a power law, and the question on the behavior of quantum effects on compression is generally open.

Chapter 5

The second order phase transitions and critical phenomena

The problem of continuous phase transitions or second order phase transitions appeared for the first time in connection with studies of the alloys ordering and phenomena associated with the orientational order. A typical example of a second-order phase transition is a transformation in the equimolar copper and zinc alloy CuZn. As can be seen in Fig. 38, the heat capacity of this alloy reveals a strong anomaly whose shape resembles the Greek letter λ. For this reason, the corresponding phase transitions are often called λ-transitions.

According to the X-ray structural analysis, the crystal structure of CuZn at low temperatures can be described as a system of primitive cubes consisting of copper and zinc atoms and inserted one into the other, as shown in Fig. 39(a). As the temperature rises, the Cu and Zn atoms begin to replace each other, and, ultimately, the probability of detecting Cu and Zn atoms in the same lattice site becomes the same. It is to this moment there corresponds a peak in the heat capacity of CuZn, which marks the phase transition point. Note that with complete mixing of the Cu and Zn atoms in the crystal structure, a

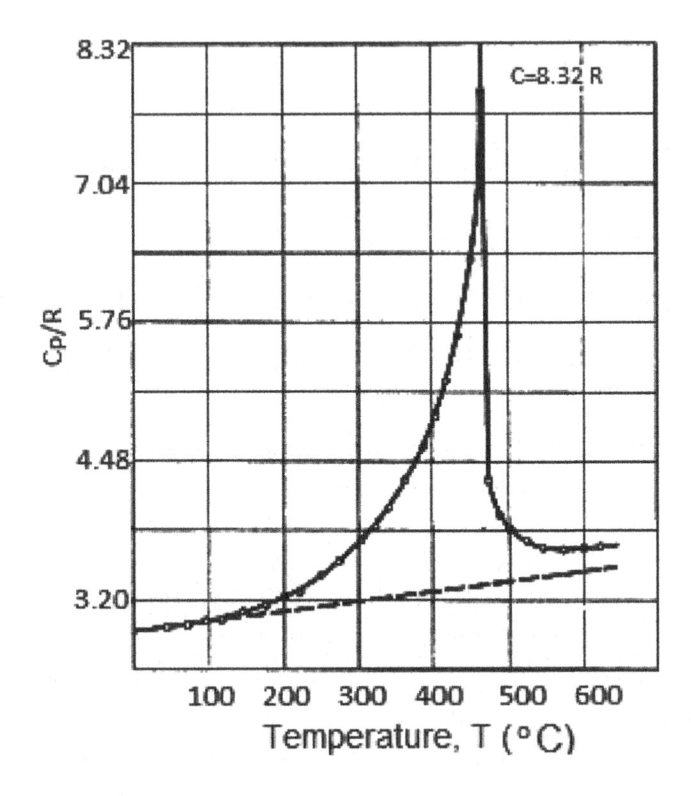

Figure 38: Heat capacity of the equimolar alloy CuZn [45].

new symmetry element (translation from the vertex to the center of the cell) appears and the crystal structure becomes volume-centered (see Fig. 39(b)).

We note that the study of the phase transformation in CuZn and in a number of other alloys is facilitated by the formation of superstructure reflexes in passing through the phase transition point, the intensity of which is directly related to the value of the "order parameter" (see Section 5.2).

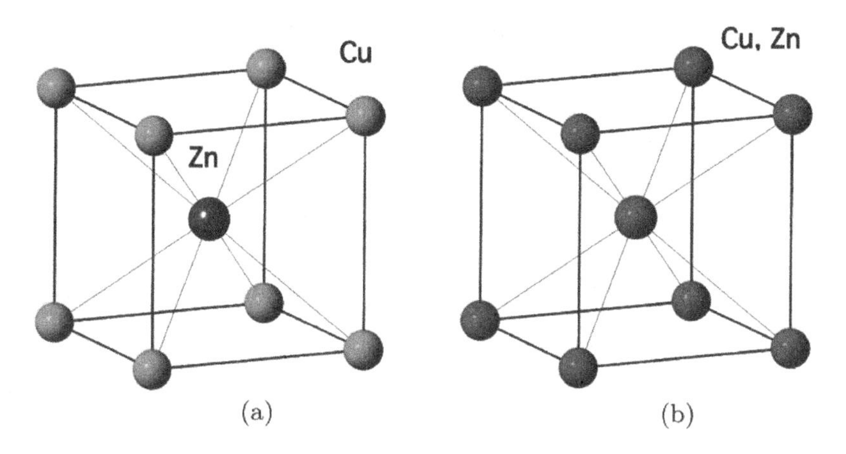

Figure 39: Elementary cells of ordered (a) and disordered phases (b) of alloy CuZn.

5.1. Bragg–Williams Theory

One of the first, or at least the most cited theories describing the ordering effects in alloys, is the Bragg–Williams theory. Let us consider this theory, mainly following Mott and Jones [46]. Suppose that we have an alloy of composition AB with total number of atoms N. Then, if the alloy is completely ordered, $N/2$ atoms of A and $N/2$ atoms of B are in their proper places in the crystal lattice, as shown in Fig. 39(a). In this case, we will refer to the position occupied by the atoms A, a-position and, accordingly, the position occupied by the atoms B, b-position. Atoms A and B in right positions a and b will be called R-atoms, and atoms A and B in the wrong positions b and a will be called W-atoms. The relations $r = R/N$ and $w = W/N$ determine the probabilities that a certain atom is R or W, respectively. $r + w = 1$. The value

$$\eta = \frac{r - w}{r + w}, \tag{44}$$

obviously, can serve as a measure of order in the system and is called the order parameter. Indeed, when all atoms belong to the set R, then $\eta = 1$, which corresponds to the complete order. However, when all atoms belong to W, then $\eta = -1$, which also corresponds to the full order. A complete disorder occurs when the probabilities r and w are equal, with the order parameter $\eta = 0$. The order defined by expression (44), obviously, should be called long-range order, since its calculation requires the involvement of all atoms of the system.

Next, we define the energy V as the energy required to move the A atom from the position a to the position b and the atom B from the position b to the position a, resulting in the number of atoms W increasing by 2, and the number of atoms R decreasing by 2. Assuming that the R and W atoms are distributed randomly over the a and b positions, we get to increase the free energy for the described exchange:

$$\delta F = V - NkT(\delta w \ln w + \delta r \ln r), \tag{45}$$

where $\delta w = 2/N$ and $\delta r = -2/N$.

Since in the state of equilibrium $\delta F = 0$ we get:

$$\frac{w}{r} = exp\left(-\frac{V}{2kT}\right), \tag{46}$$

then using the definition of the order parameter (44), we have:

$$\eta = \tanh\left(\frac{V}{4kT}\right). \tag{47}$$

Further, assuming that $V = V_o\eta$, we obtain:

$$\eta = \tanh\left(\frac{V_o\eta}{4kT}\right). \tag{48}$$

Equation (48) cannot be solved analytically. We use the expansion $\tanh x = x - 1/3x^3$ for small x. Thus, for $\eta \neq 0$ we get:

$$\eta = \frac{V_o \eta}{4kT} - \frac{1}{3}\left(\frac{V_o \eta}{4kT},\right)^3, \tag{49}$$

from which follows for the critical temperature

$$T_c = \frac{V_o}{4k}. \tag{50}$$

It is easy to obtain from (49) for the temperature dependence of the order parameter as $T \to T_c$

$$\eta = \sqrt{3}\left(\frac{T}{T_o}\right)^{3/2}\left(\frac{T_o}{T} - 1\right)^{1/2} \approx \sqrt{3}\left(1 - \frac{T}{T_c}\right)^{1/2}. \tag{51}$$

On the other hand, for $T \to 0$, it follows from (48):

$$\eta = 1 - 2\exp\left(-\frac{2T_o}{T}\right). \tag{52}$$

The general form of the temperature dependence of the order parameter η is shown in Fig. 40(a).

To calculate the heat capacity in the phase transition region, we write the energy differential E corresponding to the displacement of two atoms from the wrong ones to the correct positions in the form:

$$dE = -\frac{1}{4}NV d\eta = -\frac{1}{4}NV_o \eta d\eta. \tag{53}$$

Accordingly, the excess heat capacity resulting from the disordering, taking into account the relation (50), is:

$$\Delta c_V = \frac{1}{N}\frac{dE}{dT} = -kT_c \eta \frac{d\eta}{dT}. \tag{54}$$

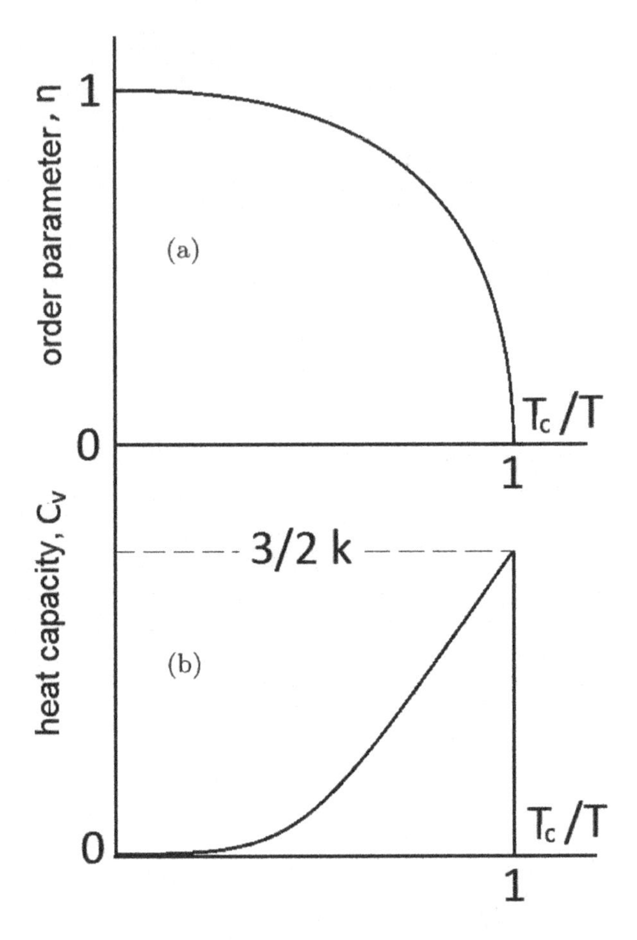

Figure 40: Behavior of (a) order parameter and (b) heat capacity at the phase transition in the Bragg–Williams model.

Differentiating (51) with respect to temperature and substituting the result in (54), we obtain the jump in specific heat at the phase transition, equal to 3/2 k per atom. Using relations (51) and (53), we obtain the dependence of excess heat capacity on the temperature in the low-temperature limit:

$$\Delta c_V = 4k\left(\frac{T_c}{T}\right)^2 \exp\left(-\frac{2T_c}{T}\right). \tag{55}$$

Relations (50), (53) and (54) allow us to represent the behavior of the heat capacity of an alloy of composition AB as shown in Fig. 40(b). Note that for $\eta = 0$, the probability of finding the atoms A and B in any of the two available positions is the same and equal to 1/2 and, consequently, the total entropy of the disorder is $S = R \ln 2$. We emphasize that the Bragg–Williams approximation is easily adapted for the analysis of any lattice systems whose nodes are occupied by particles in one of two possible states, for example, the Ising model. The comparison of Figs. 38 and 40(b) clearly shows that the Bragg–Williams theory describes a second-order phase transition in various systems only approximately and is not able to display all the details of the observed phenomena. In particular, the finite jump in the specific heat at T_c in theory does not agree with the curve in Fig. 38, which indicates the possible existence of a singularity in the behavior of specific heat for a second-order phase transition. The Bragg–Williams theory does not also describe the behavior of specific heat at $T > T_c$. The reasons for this situation are quite obvious and related to the fact that the Bragg–Williams theory, as well as the van der Waals theory and the Weiss model in the theory of magnetism, belong to the so called molecular or mean field theories. In these theories, the real interaction between particles is replaced by the interaction of particles with some average field created by the particles themselves. In the Bragg–Williams theory, the field is $V = V_o\eta$, in the van der Waals theory, $a\rho$ (ρ is the density), in the Weiss magnetism model $B = \lambda M$, where λ is the molecular field constant, M is the magnetization characterizing the long-range order in magnetic systems. Let us pay attention to the complete analogy of the formulas determining the mean field in the Bragg–Williams and Weiss models. Thus, all the effects associated with fluctuations are lost in the mean-field approximations, including the divergence of a number of thermodynamic quantities at the phase transition point.

5.2. Landau theory

5.2.1. *Phase transition of the second order*

The theory of Bragg–Williams was published in 1935, and in the same year L.D. Landau published the first version of his famous phenomenological theory of phase transitions [47] (see the complete exposition of the Landau theory in [5]). The basis of the Landau theory lies in the statement that a change of symmetry of body at a second-order phase transition is the main factor that determined properties of a phase transition. We saw above in the example of CuZn, that the symmetry of one of the phases is higher with respect to the another. In other words, the symmetric phase contains all the symmetry elements of the less symmetrical phase plus additional elements. In case of CuZn, the symmetrical phase contains an additional element of symmetry — translation from the vertex to the center of the cube. To characterize the structure in the Landau theory, the order parameter η is used, with which we have already become acquainted. As an order parameter, various quantities may appear, depending on the nature of the phase transitions under study. For the case of ordering in alloys, this can be a quantity related to the probabilities of finding at some site of the lattice of atoms A or B, in the case of phase transitions of the displacement type, the magnitude of the deviation of atoms from the position occupied in the symmetric phase. Magnetization can serve as an order parameter for magnetic phase transitions. The order parameter is defined so that it equals zero in the symmetric phase. Then in the vicinity of the phase transition point the thermodynamic potential can be expanded in a series in powers of η:

$$\Phi(P,T,\eta) = \Phi_o(P,T) + \alpha\eta + A(P,T)\eta^2 + C(P,T)\eta^3 + B(P,T)\eta^4 + \cdots . \quad (56)$$

Here $\alpha \equiv 0$ in the absence of an external field conjugate to the order parameter. The coefficient $A(P,T)$ changes its sign at

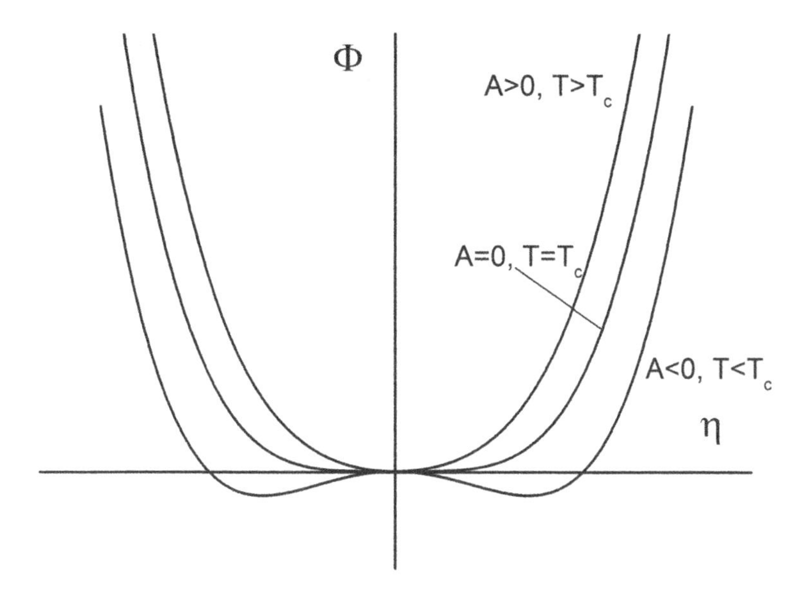

Figure 41: Behavior of thermodynamic potential in the vicinity of second-order phase transition point.

the transition point (see Fig. 41). In accordance with the stability conditions at the transition point for $\eta = 0 : A(P,T) = 0, C(P,T) = 0, B(P,T) > 0$. B is also positive in the vicinity of the transition point.

The third-order term $C(P,T)$ can be identically equal to zero, as it occurs, for example, in the case of magnetic phase transitions due to symmetry with respect to the time reversal. Then there is only one equation $A(P,T)$, which determines the existence of a line of phase transitions of the second order in the P–T plane. If $C(P,T)$ does not vanish identically, then there are only isolated points of phase transition of the second order, determined from two equations: $A(P,T) = 0$ and $C(P,T) = 0$. In the case when the third-order term $C(P,T) = 0$ is finite everywhere, then we always have a first-order phase transition. Such cases include, for example, crystallization and orientation transition of a nematic liquid crystal into an isotropic liquid. All this we will consider later, and now we turn to the canonical case when $C(P,T) \equiv 0$. The expansion of the thermodynamic potential (56) should be written in the form:

$$\Phi(P,T,\eta) = \Phi_o(P,T) + A(P,T)\eta^2 + B(P,T)\eta^4 + \cdots . \qquad (57)$$

Differentiating (57), we have:

$$\begin{aligned} \frac{\partial \Phi}{\partial \eta} = 2A\eta + 4B\eta^3 &= 0 \\ A\eta + 2B\eta^3 &= 0 \\ \eta(a + 2B\eta^2) &= 0. \end{aligned} \qquad (58)$$

Further from (58) we obtain the dependence of the order parameter on the temperature:

$$\eta^2 = -\frac{A}{2B} = \frac{a}{2B}(T_c - T). \qquad (59)$$

We write the expression for entropy in the form:

$$S = -\frac{\partial \Phi}{\partial T} = S_o - \frac{\partial A}{\partial T}\eta^2, \qquad (60)$$

from which we obtain for the heat capacity:

$$C_P = T\left(\frac{\partial S}{\partial T}\right)_P = C_{Po} + \frac{a^2 T_c}{2B}. \qquad (61)$$

Consequently, within the framework of the Landau theory, the heat capacity undergoes a simple jump at the second order phase transitions. The other quantities: compressibility, coefficient of thermal expansion, etc. also experience jumps. It should be emphasized that the behavior of the order parameter and the specific heat at the phase transition, following from the Landau theory, completely agree with the results of the van der Waals and Bragg–Williams theories, which

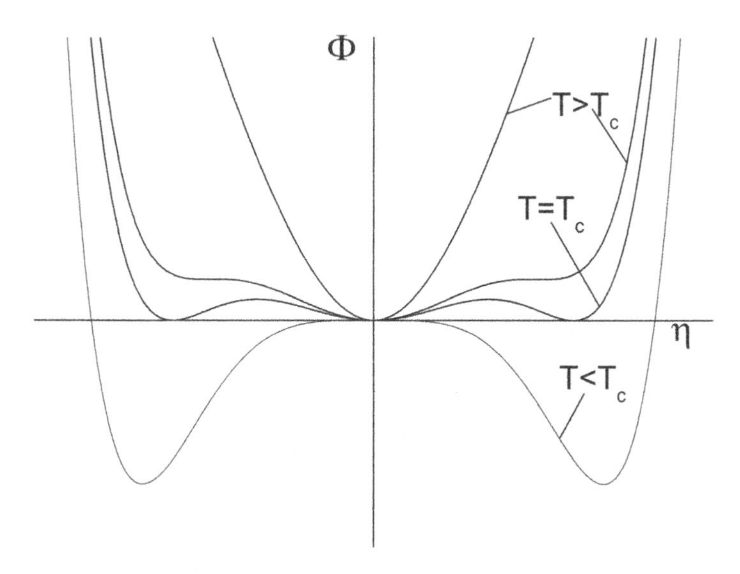

Figure 42: Thermodynamic potential for the case $A = 0, B < 0, D > 0$, which demonstrates first-order phase transition.

emphasizes the generality of mean-field theories that do not take into account fluctuations of the order parameter. Nevertheless, we also note that the behavior of the heat capacity and the coefficient of thermal expansion of tin in the region of the superconducting phase transition (see Figs. 6 and 7) are in excellent agreement with the predictions of the Landau theory. However, in other cases, the agreement is less convincing (see Fig. 38). The reasons for this situation will be clear from the following presentation.

5.2.2. *Tricritical point (critical point of continuous transition)*

Landau noted that a line of second order phase transitions can not simply terminate at a certain point, but it can go over to a line of first-order phase transitions when the sign of the coefficient B changes (see Fig. 42). The point at which this occurs, Landau called the critical

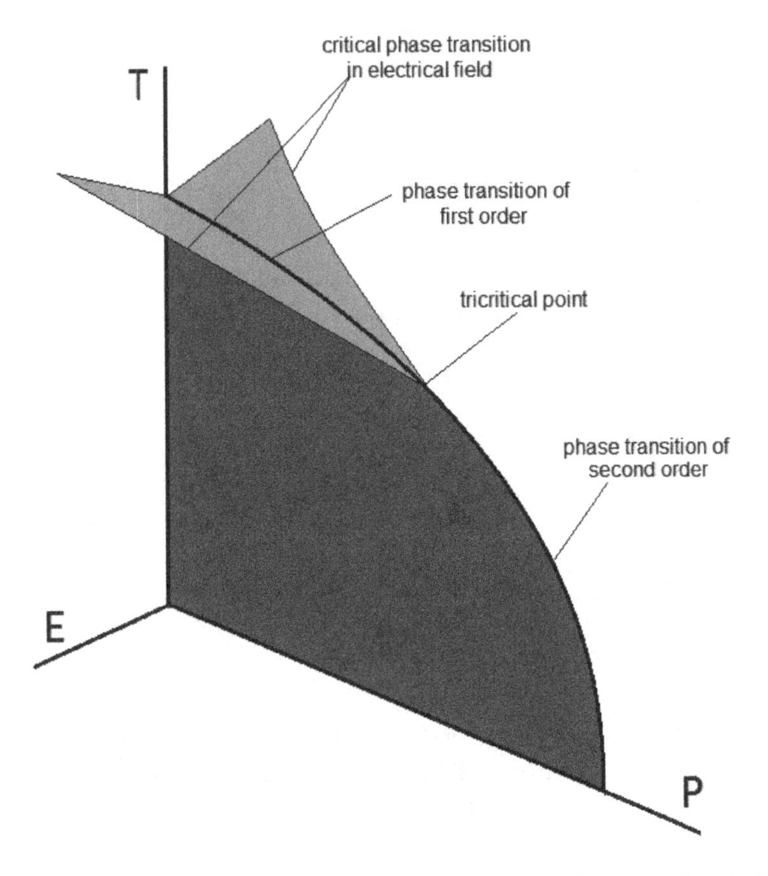

Figure 43: Schematic view of the phase diagram of ferroelectric in electric field. It is seen that at the tricritical point three lines of continuous phase transitions meet.

point of the second-order transitions. In modern literature, this point, at the suggestion of Griffiths, is called tricritical [48].

The nature of this term becomes clear from the consideration of Fig. 43, where a schematic view of the phase diagram of a ferroelectric in an electric field is demonstrated. The electric field polarizes the symmetric phase of the ferroelectric and eliminates the difference in the symmetry of both phases, thereby suppressing the second-order phase transition. At the same time, the first-order phase transition continues to exist in the electric field, vanishing only at a certain intensity as a critical phase transition. These first-order phase transition regions form wings of a kind with the opposite direction of polarization. The edges of these wings correspond to a critical phase transition, where the discontinuities of the order parameter and thermodynamic quantities vanish (see Fig. 43). As a result, three lines of continuous phase transitions are encountered at the tricritical point. To analyze this situation, we must add to the expansion (57) a sixth-order term $D(P,T)\eta^6$. The tricritical point is determined by the vanishing of the coefficients $A(P,T)$ and $B(P,T)$. For $A = 0$ and $B > 0$, the transition is continuous, but if the sign of B changes, that is, for $B < 0$, the transition becomes first-order (see Fig. 42). For a stable state of the body at the tricritical point, the sixth-order term in the expansion (57) must be retained. Of course it must be positive. So at the tricritical point, we have $A = 0,\ B = 0,\ D > 0$.

How all this happens, is depicted in Fig. 44, where the dependence of length of ferroelectric SbSJ sample on the temperature in the region of phase transition at different pressures [49] is shown. Figure 44 shows that the discontinuities of sample length and the hysteresis characterizing the first-order phase transition disappear at a tricritical point of about 1.430 kbar.

The order parameter and the heat capacity C_p at the tricritical point have the form [47]:

$$\eta^2 = \frac{1}{3D}\left[-B + \sqrt{B^2 - 3AD}\right] \tag{62}$$

or for the direction $B = 0$,

$$\eta = \left[-\frac{a(T - T_{tr})}{3D}\right]^{1/4}. \tag{63}$$

$$C_P = \left(\frac{T^2 a^3}{12D}\right)^{1/2} (T_{tr} - T)^{-1/2}. \tag{64}$$

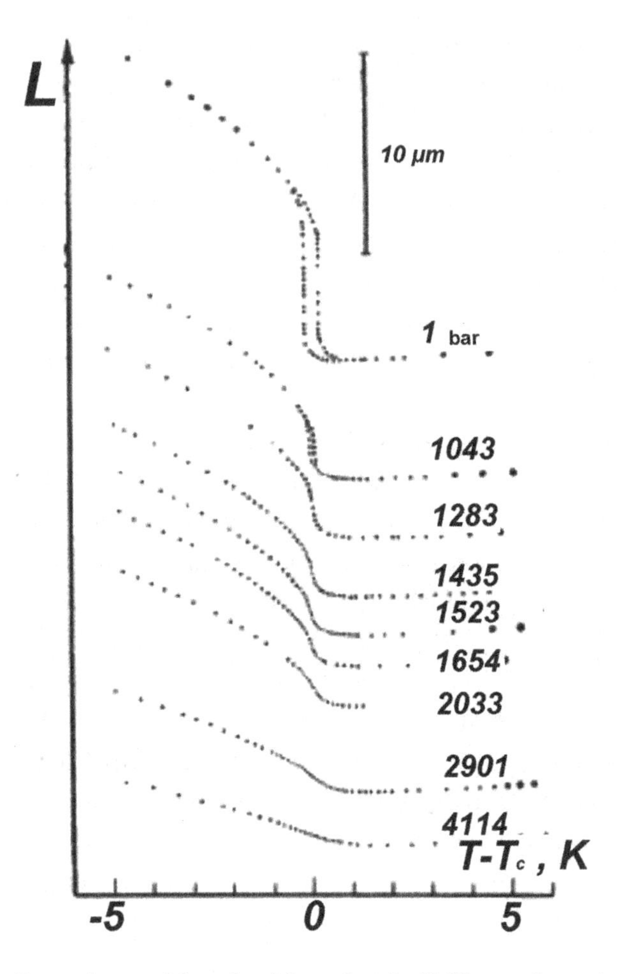

Figure 44: Dependence of length of ferroelectric SbSJ sample on temperature and pressure at the phase transitions [49].

Thus, the behavior of the order parameter and the specific heat at the tricritical point is significantly different from that of the second-order phase transition line. This is especially true for the specific heat, where a divergence arises instead of a simple jump. It is not difficult to guess that the reason for this behavior is purely geometric. Indeed, it is hardly possible to perform conjugation of a simple jump and a δ function, which characterizes the behavior of the heat capacity on either side of the tricritical point. As follows from the fluctuation theory, fluctuations play a smaller role at the tricritical point than in the case of phase transitions of the second order. For this reason, we can expect that the experimental power exponents describing the behavior of physical quantities at the tricritical point will be close to the mean-field values obtained, for example, within the framework of the Landau theory.

Let us now turn to the study of the order parameter in the ferroelectric $KH_2\,PO_4$ (KDP), carried out by scattering of γ-quantum and X-rays [50]. The fact is that although the order parameter in the case of a ferroelectric is spontaneous polarization, its direct measurement presents certain difficulties. In the case of KDP, a spontaneous deformation u_{xy} was measured that directly related to the polarization $u_{xy} = b_{36}P_z$, where b_{36} is the piezoelectric constant. The corresponding measurements were possible due to the special domain structure of the ferroelectric KDP phase (Fig. 45). By measuring the angle between two reflections using a diffractometer (see Fig. 45), it is possible to obtain the doubled value of shear angle u_{xy}. Figure 46 shows the temperature dependence of the shear

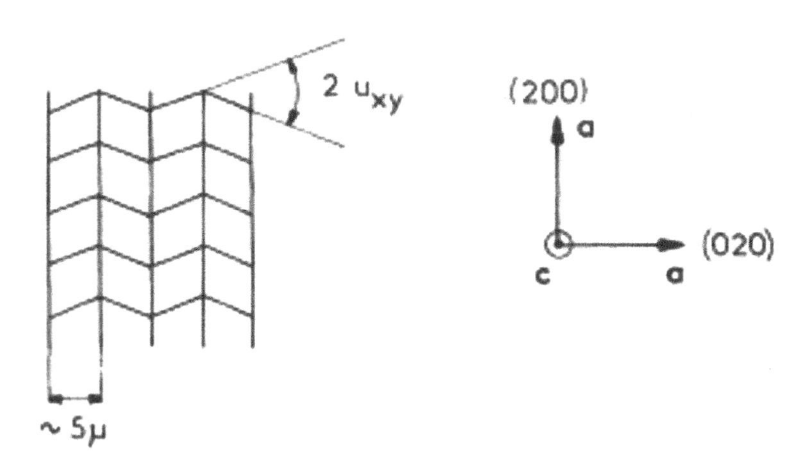

Figure 45: Domain structure of ferroelectric phase of KDP, u_{xy} is shear angle of spontaneous deformation [50].

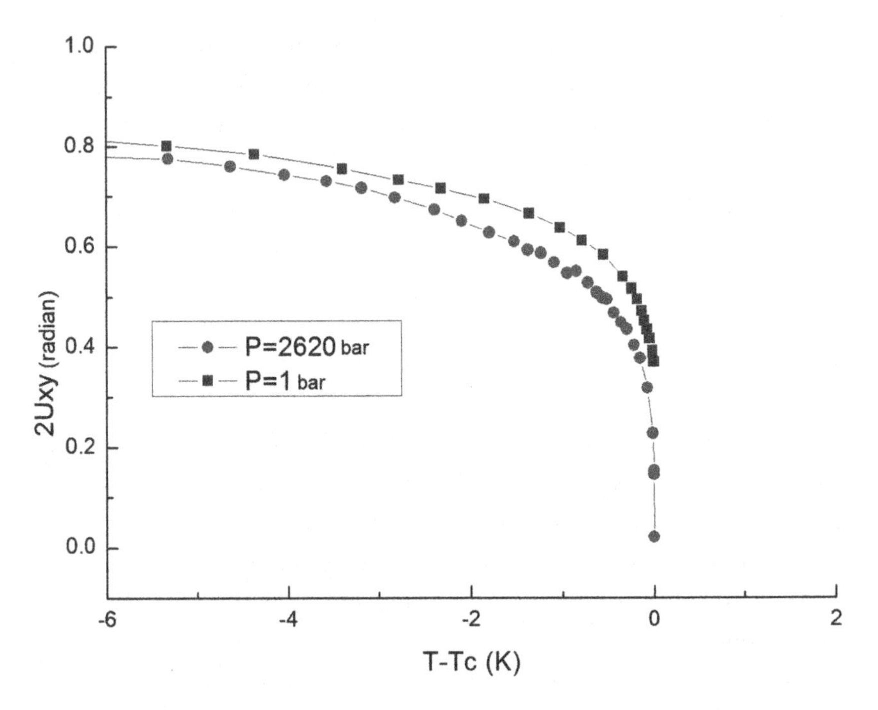

Figure 46: Dependence of shear angle u_{xy} of spontaneous deformation in KDP on temperature [51].

angle u_{xy} proportional to the order parameter P_z for spontaneous deformation in KDP [51].

Using the experiments partially illustrated in Figs. 45 and 46, the following results were obtained for the exponent β, which determines the behavior of the order parameter η for KDP near T_{tr}: $\beta = 0.225 \pm 0.03$, and $\beta = 0.235 \pm 0.02$ for the isobaric and isothermal experiments, respectively, which almost coincides with the mean-field value of $\beta = 0.25$ [52]. It should be added that from the temperature dependence of the sample length SbSJ and KDP (see Figs. 44 and 47) it is possible to determine the value of the exponent α, which determines the behavior of the specific heat, compressibility

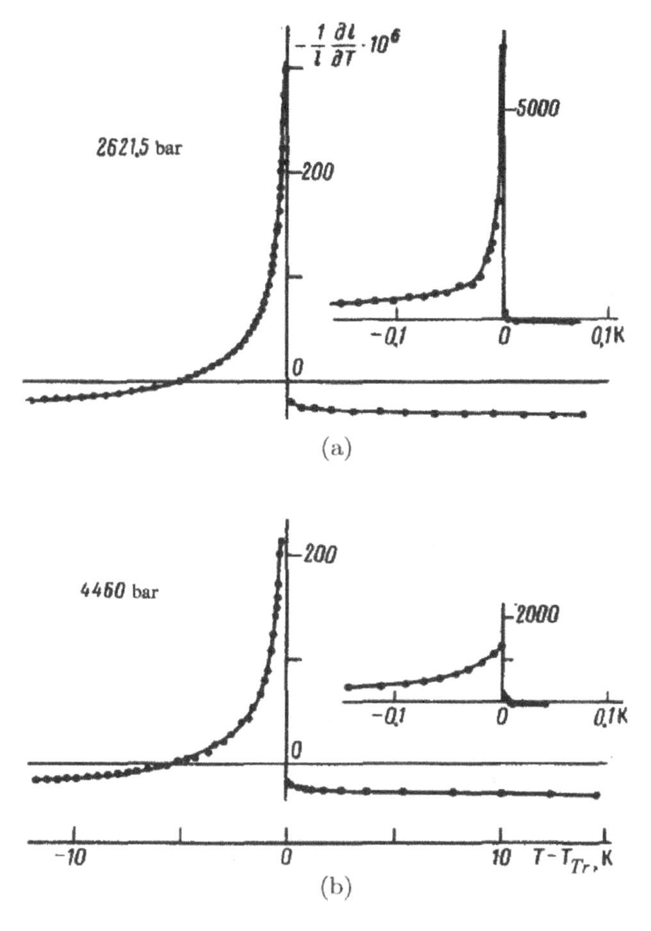

Figure 47: Temperature dependences of thermal expansion coefficient KDP in the vicinity of the (a) tricritical point and (b) second-order phase transition [52].

and thermal expansion of the ferroelectric phase at the tricritical point. So, in the case of KDP $\alpha = 0.52 \pm 0.02$, and 0.504 ± 0.03, respectively, for isobaric and isothermal experiments [52]. For SbSJ $\alpha = 0.53$ and 0.55 with a confidence interval $\delta\alpha \approx 0.05$ [49].

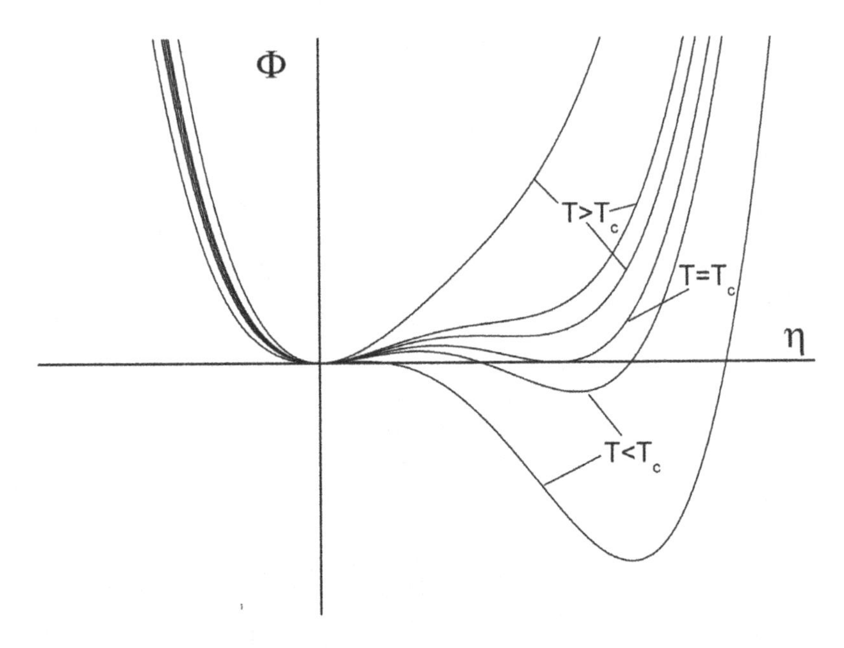

Figure 48: The thermodynamic potential when the third-order term is not zero, $C(P,T) \neq 0$.

5.2.3. *Isolated critical point*

In the case when the third-order term in (23) is not identically zero, a first-order phase transition occurs (see Fig. 48).

$$\Phi(P,T,\eta) = \Phi_o(P,T) + A(P,T)\eta^2 + C(P,T)\eta^3 + B(P,T)\eta^4 + \cdots . \tag{65}$$

However, the third-order term may be zero when changing external conditions (e.g., pressure changes) that determines a possibility of the existence of an isolated critical point, the coordinates of which are defined by simultaneously solving two equations $A(P,T) = 0$ and $C(P,T) = 0$.

A typical example of a phase transition with a nonvanishing third-order term in the Landau expansion is the nematic liquid crystal-isotropic liquid. A nematic liquid crystal is a system consisting of long molecules that can be represented in the first approximation in the form of solid rods. At low temperatures, the long axes of the molecules are directed predominantly in one direction, while at high temperatures the molecular axes are directed chaotically. The transition between these two states occurs as a first-order phase transition. The order parameter for the nematic phase transition in the simplest case can be written in the form:

$$\eta = \frac{1}{2}\langle(3\cos^2\Theta - 1)\rangle, \tag{66}$$

where Θ is the angle between the long axis of the molecule and the direction of the preferential orientation or the director. If $\Theta = 0$ and $\Theta = \pi$, then $\cos\Theta = \pm 1$ and $\eta = 1$. On the other hand, if $\Theta = \pi/2$, then $\eta = -1/2$. The situation with $\Theta = \pi/2$ corresponds to the perpendicular arrangement of the molecules with respect to the director. Accordingly, states with positive and negative order parameters are not equivalent. It is this circumstance which forces us to hold a term of the form η^3 in the expansion (56). Theoretical analysis of the nematic phase transition, held in [53], led to two hypothetical phase diagrams shown in Fig. 49. However, experimental data do not yet confirm the existence of phase diagrams of the type presented in Fig. 49. Moreover, experimental studies of the phase diagram of the liquid paroxyanisole (PAA) [54] (see Figs. 50 and 51) do not reveal any tendency to the critical behavior.

5.3. Fluctuation effects in phase transitions

5.3.1. *The Ornstein–Zernike theory [56–58]*

The critical opalescence observed at the critical phase transitions is associated with strong density fluctuations as the critical point is approached. As is known, the intensity of scattering of electromagnetic radiation by a many-particle system is determined by the

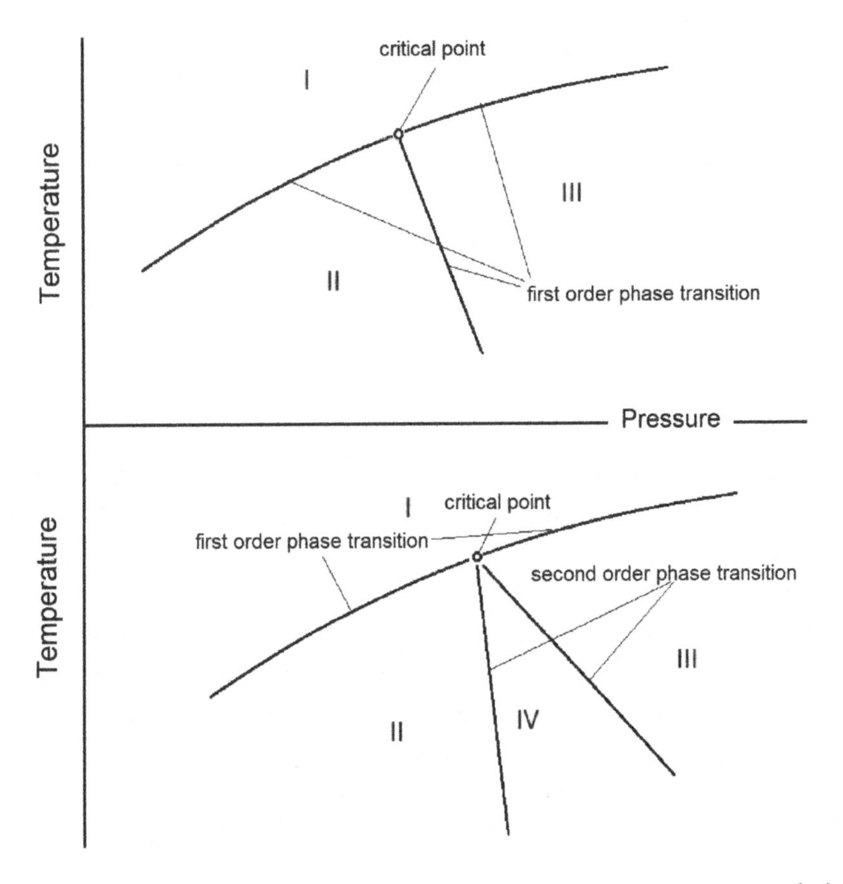

Figure 49: Two versions of phase diagrams for nematic phase transition [53]. Here I is isotropic phase, II and III are nematic phases with $\eta > 0$ and $\eta < 0$, IV — exotic "biaxial" phase.

expression:

$$\frac{I(q)}{I^0(q)} = 1 + \rho \int [g(r) - 1] e^{-iqr} dr = S(q), \tag{67}$$

where $I^0(q)$ is the scattering by a system of noninteracting particles, ρ is the density, $g(r)$ is the radial distribution function, and $S(q)$ is the structure factor.

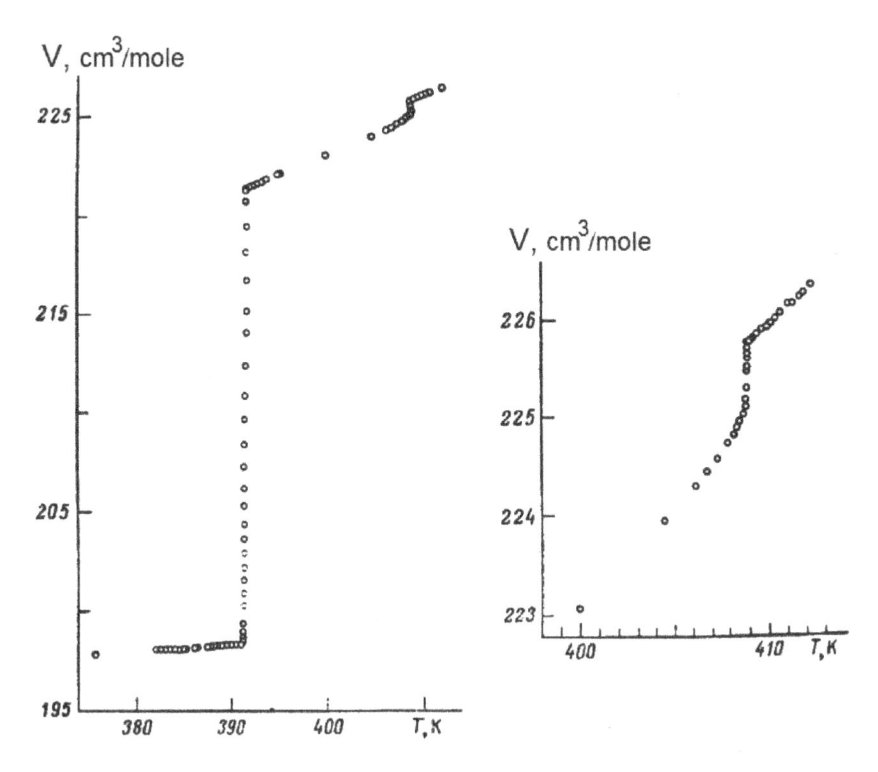

Figure 50: Dependence of molar volume of liquid crystal substance — para — oxyanisole (PAA) on temperature at melting and in nematic phase transition. On the right, the region of nematic phase transition is shown on enlarged scale [55].

Since the function $g(r)$ oscillates around unity, it is convenient to replace it in the sequel with the function $h(r)$:

$$h(r) = g(r) - 1. \tag{68}$$

The Ornstein–Zernike theory is based on the representation of the function $h(r)$ in the form:

$$h(r_{12}) = c(r_{12}) + \rho \int c(r_{13}) h(r_{32}) dr_3, \tag{69}$$

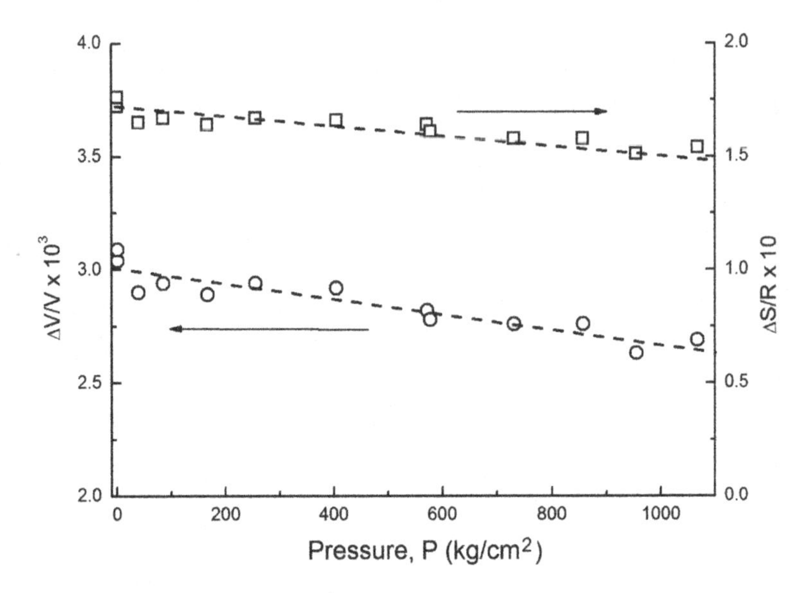

Figure 51: Relative change of volume and change of entropy at the first-order phase transition in PAA at high pressures [53].

where $c(r)$ is the so-called direct correlation function. The Fourier transform (69) leads to the relation:

$$\hat{h}(q) = \hat{c}(q) + \rho\hat{c}(q)\hat{h}(q). \tag{70}$$

It is assumed that the Fourier transform of the direct correlation function is an even analytic function and can be expanded in a Taylor series in a neighborhood of $q = 0$. By breaking the series on the quadratic term, we have:

$$\hat{c}(q) = \hat{c}(0) - L^2q^2 + O(q^4). \tag{71}$$

Further, using (70) and (71) we obtain:

$$\hat{c}(q) = \frac{\hat{h}(q)}{1 + \rho\hat{h}(q)} = \frac{S(q) - 1}{S(q)}, \tag{72}$$

$$S(q) = \frac{L^{-2}}{L^{-2}[1 - \hat{c}(0)] + q^2} \tag{73}$$

or, using the notation $\xi^{-2} = L^{-2}[1 - \hat{c}(0)]$, we have:

$$S(q) = \frac{L^{-2}}{\xi^{-2} + q^2}. \tag{74}$$

The inverse Fourier transform gives:

$$h(r) \sim L^{-2}\frac{e^{-r/\xi}}{r}, \tag{75}$$

from which it follows that ξ characterizes the correlation radius. Usually ξ is called the correlation length.

Further, using the known relation:

$$S(0) = \frac{\beta_T kT}{V}, \tag{76}$$

where β_T is the compressibility, and taking into account the dependence (20), following from the van der Waals model, we see that the correlation length diverges at the critical point according to the law:

$$\xi \propto \tau^{-1/2}, \tag{77}$$

or in the general case:

$$\xi \propto \tau^{\nu}. \tag{78}$$

Taking into account that the compressibility or susceptibility diverges at the critical point as $\beta_T \propto \tau^{-\gamma}$, where $\gamma = 1$, we obtain the ratio for two critical exponents in the Orstein–Zernike model,

$$\nu = \frac{1}{2}\gamma. \tag{79}$$

From the foregoing (see (75) and (77)) it follows that for $\tau \to 0$, $\xi \to \infty$ and $h(r) \to 1/r$. The latter means that at the critical point the correlations become extremely long-range. As can be seen from (76), $S(q)$ also diverges at the critical point for $q = 0$, which is actually the cause of strong light scattering (opalescence). We point out that in the general case of a d-dimensional space for $\tau \to 0$, $\xi \to \infty$:

$$h(r) \to r^{-(d-2)}; \quad d \geq 3, \tag{80}$$

$$h(r) \to \ln r; \quad d = 2. \tag{81}$$

The relationship (81) has no physical meaning (correlation increases with distance), which is one of the drawbacks of the Ornstein–Zernike theory. To remedy this shortcoming, M. Fisher introduced a new critical exponent η, describing the behavior of the structure factor for small wave vectors and temperatures close to critical. As a result, expression (80) takes the form:

$$h(r) \to r^{-(d-2+\eta)}; \quad r \to \infty. \tag{82}$$

5.3.2. *Fluctuations in the Landau theory*

The Landau theory in the simple version does not take into account at all the fluctuations of the order parameter and therefore is not valid in the region close to the phase transition. The simplest way to allow for fluctuations is to expand the thermodynamic potential not only in powers of the order parameter, but also in its gradients. Restricting ourselves to the first significant term of the gradient expansion and leaving only the quadratic term in the expansion (57), we write:

$$\Phi - \Phi_o = A\eta^2 + g\left(\frac{\partial \eta}{\partial x}\right)^2, \tag{83}$$

or

$$\Delta\Phi = A\left[\eta^2 + \frac{g}{A}\left(\frac{\partial\eta}{\partial x}\right)^2\right]. \tag{84}$$

Here the ratio g/A is a quantity with the dimension of squared length. Denoting g/A by ξ^2, we write:

$$\xi^2 = \frac{g}{A} = \frac{g}{a(T - T_c)}. \tag{85}$$

It follows from (85) that ξ is a correlation length characterizing the spatial inhomogeneity of the system. It is seen that ξ diverges at the critical point according to the law:

$$\xi = \xi_o(T - T_c)^{-1/2}, \tag{86}$$

which completely agrees with the Ornstein–Zernike theory. We note that in the general case the dependence (85) is written in the form:

$$\xi = \xi_o(T - T_c)^{-\nu}. \tag{87}$$

Further, taking into account the relations [5]:

$$\eta^2 = -\frac{A}{2B}; \quad \Delta\eta^2 = \frac{T_c}{AV_c} = \frac{T_c}{A\xi^d},$$

for relative mean-square fluctuations of the order parameter (see also [59, 60]):

$$\frac{\langle\Delta\eta^2\rangle}{\eta^2} = \frac{T_c^2}{\Delta C_p \xi_o^d (T - T_c)^{2-\frac{d}{2}}}; \tag{88}$$

here d is the spatial dimension.

Using relation (88), it is possible to estimate the width of the fluctuation region and the limits of applicability of the Landau theory. Expression (88) also establishes an important relationship between spatial dimension and intensity of fluctuations. It follows from (88) that for $d > 4$, the order parameter fluctuations are finite for any $\delta T = T - T_c$, respectively, the order parameter fluctuations diverge at $d < 4$. The dimension $d = 4$ in this case is the upper critical dimension d_c^+. We add that there is also a lower critical dimension d_c^-, for which long-range order does not exist in the system at any finite temperatures.

5.4. Correlation length and critical exponents

From (88) it follows that it is the seed value of the correlation length ξ_o that, under all other equal conditions, determines the width of the fluctuation region in the phase transition. For example, in the case of classical superconductors with coherence length 10^{-4} cm, the fluctuation effects are practically not observable (see Fig. 3). At the same time, high-temperature superconductors with a coherence length of 10^{-6} cm clearly demonstrate the existence of a fluctuation anomaly at the phase transition (Fig. 52). Here, the coherence length plays the role of a correlation length.

The correlation length is the most important parameter determining the critical behavior of a system. In particular, as we shall see later, the behavior of the heat capacity in the critical region is related to the correlation length by means of the relation:

$$\alpha = 2 - \nu d, \tag{89}$$

where α and ν are the exponents related to the expressions $C_v \sim (T - T_c)^{-\alpha}$ and $\xi = \xi_o(T - T_c)^{-\nu}$, d is the spatial dimension. Substituting

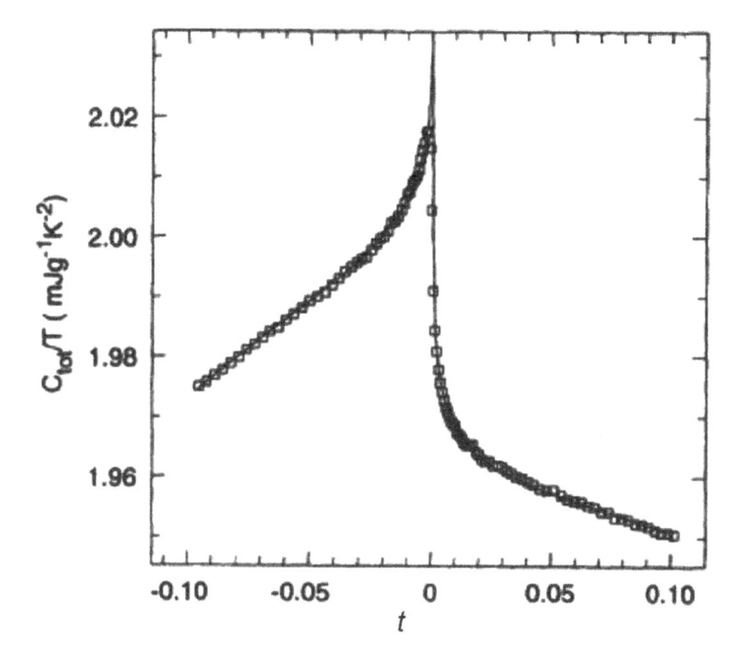

Figure 52: Heat capacity of high-temperature superconductor $YBa_2Cu_3O_7$ as a function of reduced temperature [61].

the value $\nu = 1/2$ (77) in (89), we obtain $\alpha = 1/2$ for $d = 3$. We note that the value $\alpha = 1/2$ corresponds to the so-called "Gaussian" approximation in the theory of phase transitions [62].

An example of a "Gaussian" behavior is observed in the band ferromagnet $SrReO_3$ [63] (Fig. 53). The temperature dependence of the heat capacity of this material in the region of the magnetic phase transition is consistent with the assumption $\alpha' = \alpha = 0.5$. The magnetic measurements made it possible to determine also other critical exponents for $SrReO_3$: $\beta = 0.50 \pm 0.3$, $\gamma = 0.99 \pm 0.3$, $\delta = 3.1 \pm 0.3$, which, as one would expect, practically coincide with the mean-field values. We should also point out the paper [64], which involves Gaussian fluctuations in the analysis

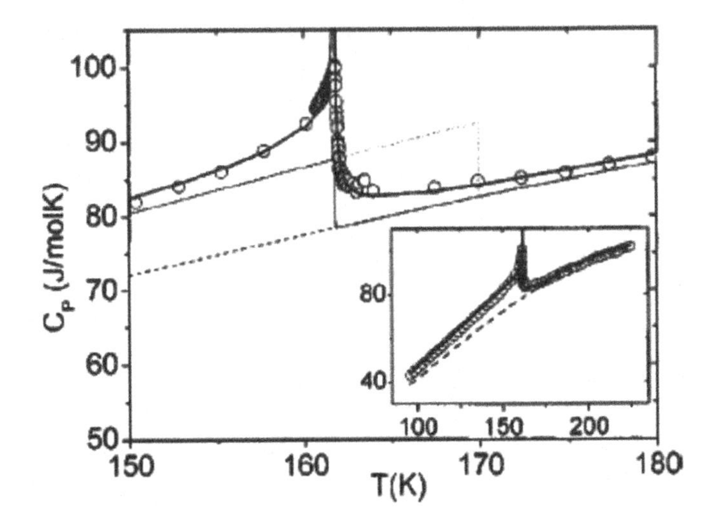

Figure 53: Specific heat of ferromagnet $SrReO_3$ in the magnetic phase transition region. Thick solid lines — approximation taking into account Gaussian fluctuations [63].

of the Young's modulus behavior in the superconducting phase transition in $YBa_2Cu_3O_7$ (Fig. 54). In fact, the Gaussian model is rarely realized in an experiment. The point is that in this kind of harmonic model, the interaction of fluctuations or the anharmonic contribution are not taken into account. However, nature is arranged in such a way that either fluctuations are almost invisible, or they are so strong that their interaction is inevitable. For this reason, experimentally measured values mostly differ from mean field and Gaussian values (see Table 1). Measurements of critical exponents played an important role in the study of phase transitions and the construction of theoretical models.

Table 1 gives the definitions of the most important critical exponents for the liquid and magnetic systems. Table 2 provides examples of critical exponents for a number of model and real systems. Figures 55–58 illustrate the behavior of heat capacity at the critical point SF_6, the superfluid phase transition in ^4He

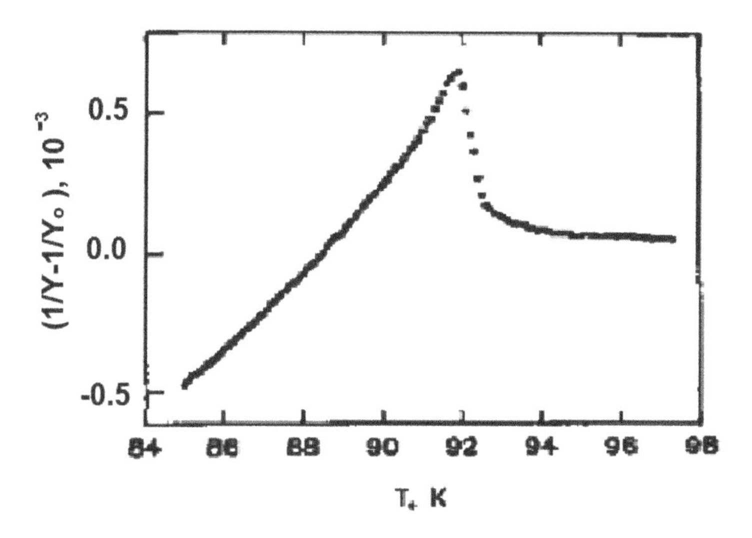

Figure 54: Behavior of reciprocal of the Young's modulus in the region of the superconducting phase transition in $YBa_2Cu_3O_7$. An analysis of the data indicates an existence of fluctuation (Gaussian) contribution to the elastic properties of the compound [64].

Table 1: Critical exponents for liquid and magnetic systems.

Exponents	Liquid Systems	Magnetic Systems	Classical Values
α'	$C_v \sim (-\tau)^{-\alpha'}$	$C_H \sim (-\tau)^{-\alpha'}$	0
α	$C_v \sim \tau^{-\alpha}$	$C_H \sim \tau^{-\alpha}$	0
β	$\rho_l - \rho_g \sim (-\tau)^{\beta}$	$M \sim (-\tau)^{\beta}$	1/2
γ'	$K_T \sim (-\tau)^{-\gamma'}$	$\chi_T \sim (-\tau)^{-\gamma'}$	1
γ	$K_T \sim \tau^{-\gamma}$	$\chi_T \sim \tau^{-\gamma}$	1
δ	$P - P_c \sim \|\rho - \rho_c\|^{\delta}$	$H \sim \|M\|^{\delta}$	3
ν'	$\xi \sim (-\tau)^{-\nu'}$	$\xi \sim (-\tau)^{-\nu'}$	1/2
ν	$\xi \sim \tau^{-\nu}$	$\xi \sim \tau^{-\nu}$	1/2
η	$G(r) \sim \|r\|^{-(d-2+\eta)}$	$G(r) \sim \|r\|^{-(d-2+\eta)}$	0

Table 2: Critical exponents for experimental and model systems.

System	α	β	γ	δ	ν	η
Mean field	0 (discontinuity)	1/2	1	3	1/2	0
Ising model (d = 2, n = 1), exact solution	0 (logarithm)	1/8	7/4	15	1	1/4
Ising model (d = 3, n = 1), approximation	0.110	0.325	1.24	4.82	0.63	0.03
Heisenberg model (d = 3, n = 3) approximation,	−0.10	0.36	1.38	4.80	0.705	0.03
X-Y model (d = 3, n = 2)	−0.0146	0.3485	1.3177	4.780	0.67155	0.038
Unixial magnetics (d = 3, n = 1)	0.08–0.1	0.33–0.35	1.15 ± 0.02			
^{4}He, λ - transition (d = 3, n = 2)	−0.0127 ± 0.0003				0.6705 ± 0.0006	
Liquids (d = 3, n = 1)						
Xe	0.08 ± 0.02	0.325–0.337	1.23 ± 0.03	4.40		
SF_6	0.1105 ± 0.025	0.321–0.339	1.25 ± 0.03		0.621	0.03 ± 0.03
Ferromagnets (d = 3, n = 3)						
Fe	−0.09 ± 0.01	0.34 ± 0.2	1.33 ± 0.02			0.07 ± 0.07
Ni	−0.09 ± 0.03	0.37 ± 0.3	1.34 ± 0.02	4.2 ± 0.1		

and the Heisenberg magnet EuS. One should pay attention to the unique temperature resolution achieved in studying the He I–II phase transition under conditions of weightlessness (Fig. 56). Initially, in the heroic era of the study of phase transitions, marked by the work of M. Buckingham and W. Fairbanks [68] on the study of the λ transition in liquid helium and the works of A.V. Voronel and co-workers [69, 70] on the measurement of the specific heat at critical points of simple liquids, it was believed that the critical exponents are universal, that is, they do not depend on the specific substance and type of phase transition. Such statements were largely based on the divergence of the correlation length at the critical point. Indeed, in the critical region, the correlation length significantly exceeds the characteristic lengths inherent in the substance (particle size, interatomic distances), which makes microscopic properties of materials unimportant. The data in Table 2 show that this assumption is not entirely justified. L. Kadanoff [69] was one of the first to realize this problem and introduced the concept of universality classes, each of which is characterized by a certain dimensionality of space, symmetry of the order parameter and the interaction radius, but not its detailed shape and amplitude (see Table 2).

Accordingly, the critical exponents of systems belonging to the same universality class should be the same. For example, phase transitions in the 3d Ising model, uniaxial magnets, liquid-gas transitions, order-disorder in alloys are characterized by the same set of critical exponents.

Here it is necessary to emphasize that an increase of the radius or length of interparticle interaction means an increase in the intensity of mean field acting on the particle. Similarly, an increase of the space dimensionality is accompanied by an increase in the number of particles within the range of interaction. Both reduce the probability of fluctuations. As a result, phase transitions in models with an infinite interaction radius are described in the framework of

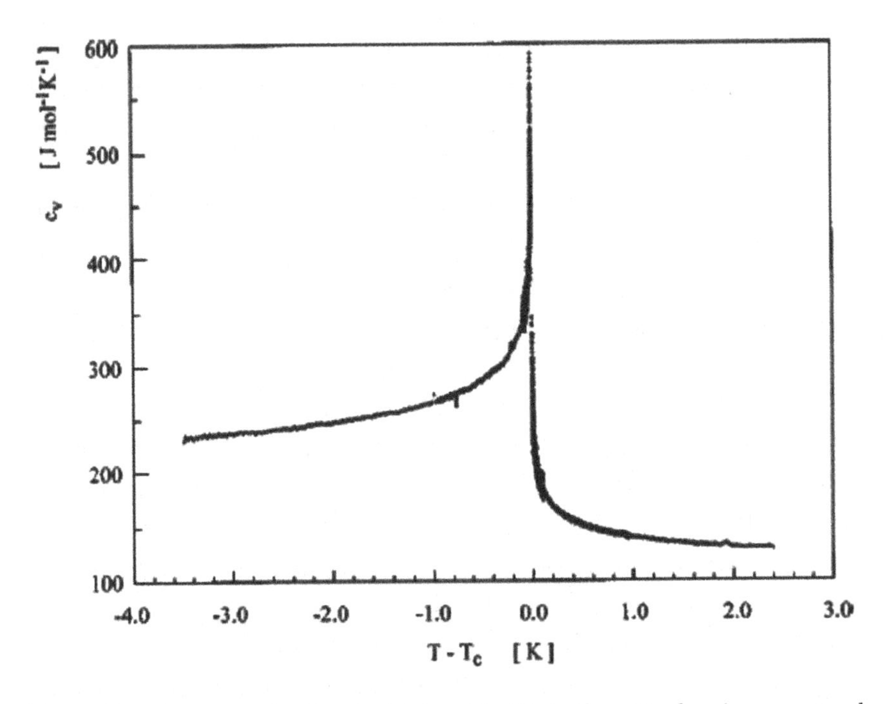

Figure 55: Heat capacity C_V of sulfo-fluoride SF_6 in the critical region, measured under zero-gravity conditions to avoid gravitational effects [65]. As a result, the value of exponent $\alpha = 0.1105$ was obtained.

mean-field theory. Models with short-range interaction in the space $d > 4$ also show a suppression of the fluctuation contribution to the thermodynamic characteristics of the phase transition. As we see, the critical exponents are not completely universal quantities. However, as it turned out (see, for example, [46,56,62,72]), there are a number of truly universal relations between the exponents, called similarity relations. These relations, listed below, bear the names of the authors who proposed them:

$$\gamma = \nu(2 - \eta) \text{ (Fisher)},$$

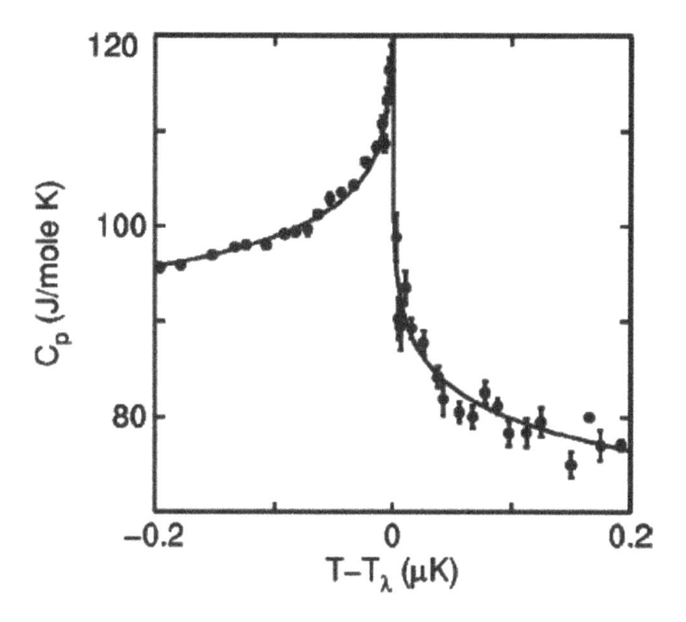

Figure 56: The specific heat ^{4}He at the superfluid phase transition He I–II [66]. Measurements were carried out under zero-gravity conditions, which made it possible to determine the exponent α with an accuracy unattainable before ($\alpha = -0.0127 \pm 0.0003$). We draw attention to the negative sign of exponent, which means that the peak of heat capacity is finite at the He I–II phase transition in accordance with the x-y model.

$$\begin{aligned} \alpha + 2\beta + \gamma = 2 \text{ (Rashbrook)}, \\ \gamma = \beta(\delta - 1) \text{ (Vidom)}, \\ \nu d = 2 - \alpha \text{ (Josephson)}. \end{aligned} \tag{90}$$

Note that only two of the six critical exponents are independent. The remaining four can be determined from relations (90).

In the present discussion, I would like to emphasize the special role of the correlation length ξ as the most important factor determining the critical properties of the system. We will derive relations (90), assuming that the correlation length ξ is the only "characteristic"

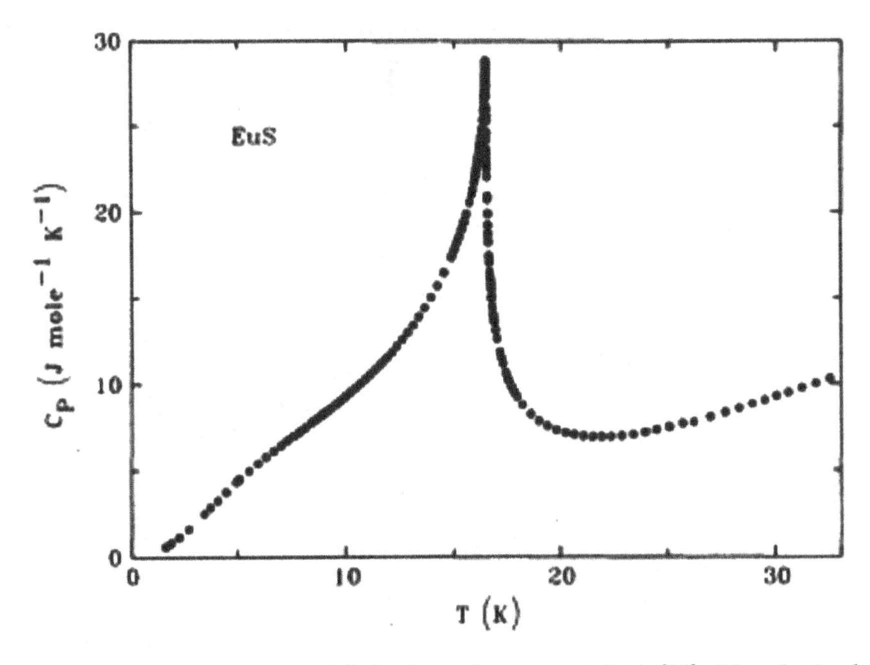

Figure 57: Heat capacity of Heisenberg ferromagnet EuS [67]. The obtained value of the specific heat exponent $\alpha = -0.13$ indicates the finite heat capacity peak at phase transition, as follows from the 3D Heisenberg model (see also Fig. 58).

length, in terms of which all other lengths of the system must be measured. Next we will use dimensional considerations.

The ratio F/kT, where F is the free energy, is dimensionless, respectively, $f = F/kTV$, where V-volume has the dimension of the inverse length

$$[f] = L^{-d} \tag{91}$$

or, using the correlation length as a measure of length, we write (91) in the form:

$$[f] = \xi^{-d}. \tag{92}$$

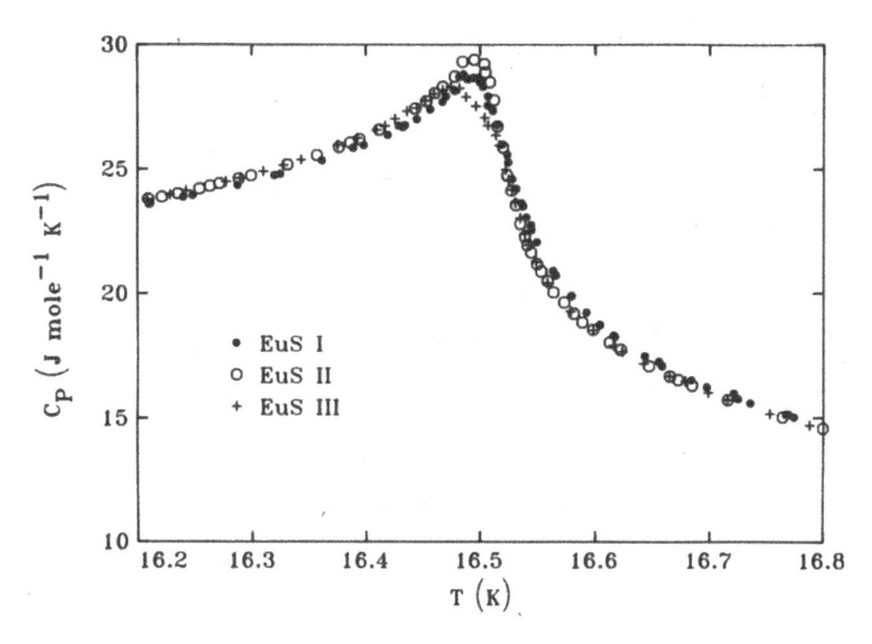

Figure 58: Heat capacity of three EuS samples in the vicinity of a phase transition in a narrow temperature interval, illustrating an absence of divergence in the 3D Heisenberg universality class [67].

Accordingly, for the heat capacity dimension, taking into account the ratio $\xi \sim \tau^{-\nu}$, we have:

$$C \sim \frac{\partial^2 f}{\partial \tau^2} \sim f\tau^{-2} \sim \xi^{-d}\tau^{-2} \sim \tau^{\nu d-2},$$

which in connection with the definition of $C \sim \tau^{-\alpha}$ immediately leads to the Josephson relation (90)

$$\nu d = 2 - \alpha. \tag{93}$$

From the definition of the order parameter follows

$$m \sim \frac{\partial f}{\partial h} \sim fh^{-l} \sim \xi^{-d}h^{-l},$$

where m is the order parameter, and h is the conjugate field. Further, using relations $m \sim h^{1/\delta}$ and $m \sim \tau^{\beta}$, we get

$$\tau^{\nu d - \delta\beta} = \tau^{\beta},$$

which leads to the expression

$$\nu d - \delta\beta = \beta, \tag{94}$$

that we use later.

For the susceptibility we write

$$\chi \sim \frac{\partial^2 f}{\partial h^2} \sim f h^{-2} \sim \xi^{-d} h^{-2} \sim \tau^{\nu d - 2\delta\beta},$$

which, taking into account the definition of $\chi_T \sim \tau^{-\gamma}$, gives

$$\nu d - 2\delta\beta = -\gamma. \tag{95}$$

The subsequent subtraction of (94) from (95) with regard to (93) leads to the relations of Widom (96) and Rashbrook (97) (see (90))

$$\gamma = \beta(\delta - 1) \tag{96}$$

and

$$\alpha + 2\beta + \gamma = 2. \tag{97}$$

Finally, we obtain the Fisher ratio (90), using the dimension of correlation function (82),

$$g(\xi) \sim \xi^{-(d-2+\eta)} \sim \tau^{\nu(d-2+\eta)}.$$

Since the dimension of correlation function is equal to the dimension of square of order parameter [5], we have:

$$\nu d = 2\nu - \eta\nu + 2\beta. \tag{98}$$

From (94), (95), and (98) the desired relation follows

$$\gamma = \nu(2 - \eta). \tag{99}$$

The similarity relations (90) can also be obtained if we assume that the singular part of the thermodynamic potential in the vicinity of the phase transition is a generalized homogeneous function of the reduced temperature and field. In fact, this assumption means that the singular part of the potential can be reduced to a form that depends only on a single variable, including a combination of temperature and field. This is what constitutes the subject of so-called scale invariance, which found the rationale within the framework of the renormalization-group approach. The same approach allowed us to calculate the numerical values of the critical exponents [57, 62, 72] for the first time. It should also be noted that the analysis of the phase transition in a system of particles interacting via a power-law potential, carried out in Section 3.2, is another example of the application of the principle of scale invariance.

On this I will stop my presentation, intended for beginners. To continue education, one should refer to the special literature on phase transitions.

Bibliography

[1] I.N. Makarenko, V.A. Ivanov, S.M. Stishov, Pribory i Tekhnika Eksperimenta, **3**, 202 (1974).

[2] I.N. Makarenko, A.M. Nikolaenko, V.A. Ivanov, S.M. Stishov, ZhETF, **69**, 1723 (1975) (English translation: I.N. Makarenko, A.M. Nikolaenko, V.A. Ivanov, S.M. Stishov, JETP, **42**, 875 (1975)).

[3] D. Schoenberg, "Superconductivity" Cambridge University Press, Cambridge (1952).

[4] G.K. White, Phys. Lett. **8**, 294 (1964).

[5] L.D. Landau, E.M. Lifshits, "Statistical physics" Pt.1 (Pergamon Press, Oxford 1980).

[6] H.W. Habgood, W.G. Schneider, Can. J. Chem. **32 (2)**, 98 (1954).

[7] Yu.B. Rumer, M.Sh. Ryvkin, "Thermodynamics, Statistical Physics and Kinetics" (Mir, Moscow, 1980).

[8] E.A. Guggenheim, J. Chem. Phys. **13**, 253 (1945).

[9] G.L. Pollack, Rev. Mod. Phys. **36**, 748 (1964).

[10] F. Hensel, M. Stolz, G. Hohl, R. Winter and W. Gotzzlaff, Journal de Physique IV, Colloque C5, supplement au Journal de Physique I, **1**, decembre, 5-191 (1991).

[11] L.D. Landau, Ya. B. Zel'dovich, ZhETF, **14**, 32, 1944.

[12] J.C. Hensel, T.G. Phillips, G.A. Thomas, The electron-hole liquid in Semiconductors, in Solid State Physics, vol. 32, Eds. H. Ehrenreich, F. Seitz, D. Turnbull, Academic Press (1977).
[13] S.M. Stishov, ZhETF Pis'ma, **57**, 189 (1993) (English translation: S.M. Stishov, JETP Letters, **57**, 196 (1993)).
[14] M.Y.J. Hagen, E.J. Meijer, G.C.A.M. Mooij, D. Frenkel, H.N.W. Lekkerkerker, Nature, **365**, 425 (1993).
[15] C.F. Tejero, A. Daanoun, H.N.W. Lekkerkerker, M. Baus, Phys. Rev. Lett. **73**, 752 (1994).
[16] L.S. Ornstein, and F. Zernike, Proc. Acad. Sci. Amsterdam, **17**, 793 (1914).
[17] F.Datchi, P.Loubeyre, R. LeToullec, Phys. Rev. **B61**, 6535 (2000).
[18] L. Landau, ZhETF, **7**, 19 (1937).
[19] T. Hill "Statistical Mechanics" (New York, MacGraw Hill, 1956).
[20] W.W. Wood, J.D. Jacobson, J. Chem. Phys. **27**, 1207 (1957).
[21] B.J. Alder, T.E. Wainwright, J. Chem. Phys. **27**, 1208 (1957).
[22] B.J. Alder, T.E. Wainwright, Phys. Rev. **127**, 359 (1962).
[23] L.K. Runnels, J. Chem. Edu. **47**, 742 (1970).
[24] S.M. Stishov, UFN, **114**, 3 (1974). (English translation: S.M. Stishov, Sov. Phys. Usp. **18**, 625 (1975)).
[25] I.Z. Fisher, "Statistical theory of liquids" (The University of Chicago Press, 1964).
[26] D.A. Young, "Phase diagram of Elements" (Univ. of California Press, 1991).
[27] W.G. Hoover, M. Ross, Contemp. Phys., **12**, 339 (1971).
[28] J.O. Hirschfelder, D.P. Stevenson, H. Eyring, J. Chem. Phys., **5**, 896 (1937).
[29] S.M. Stishov, Phil. Mag. **82**, 1287 (2002).
[30] D.A. Young, B.J. Alder, Phys. Rev. Lett. **38**, 1213 (1977).
[31] E. Brindeau, R. Levant, J-P. Hansen, Phys. Lett. **60A**, 424 (1977).
[32] A. Jayaraman, Phys. Rev. **137**, A 179 (1965).
[33] I.N. Makarenko, V.A. Ivanov, S.M. Stishov, ZhETF Pis'ma, **18**, 320 (1973) (English translation: I.N. Makarenko, V.A. Ivanov, S.M. Stishov, JETP Letters, **18**, 187 (1973)).

[34] S.M. Stishov, UFN, **96**, 467 (1968) (English translation: S.M. Stishov, Sov. Phys. Usp. **11**, 816 (1969)).

[35] S.M. Stishov, UFN, **154**, 93 (1988) (English translation: S.M. Stishov, Sov. Phys. Usp. **31**, 52 (1988)).

[36] R.H. Sherman, in the collection of "Physics of the High Pressres and Condensed Phase" Edited by A. van Itterbeek (North-Holland Publishing Company — Amsterdam, 1965).

[37] I. Pomeranchuk, ZhETF, **20**, 919 (1950).

[38] D.A. Kirzhnits, ZhETF, **38**, 503 (1960) (English translation: D.A. Kirzhnits, JETP **11**, 365 (1960)).

[39] D.A. Kirzhnits, UFN, **104**, 489 (1971) (English translation: D.A. Kirzhnits, Sov. Phys. Usp. **14**, 512 (1972)).

[40] A.A. Abrikosov, ZhETF, **39**, 1800 (1960) (English translation: A.A. Abrikosov, JETP, **39**, 1800 (1960)).

[41] G. Chabrier, N.W. Ashcroft, H.E. DeWitt, Nature, **360**, 48 (1992).

[42] M.D. Jones, D.M. Ceperley, Phys. Rev. Lett. **76**, 4572 (1996).

[43] F.A. Lindemann, Phys. Z., **11**, 609 (1910).

[44] S.M. Stishov, UFN, **171**, 299 (2001) (English translation: S.M. Stishov, Phys. Usp. **44**, 285 (2001)).

[45] F.C. Nix, W. Shockley, Rev. Mod. Phys. **10**, 2 (1938).

[46] N.F. Mott, H. Jones, The theory of the properties of metals and alloys, (Oxford University Press, 1936).

[47] L.D. Landau, Phys. Zs. Sowjet. **8**, 113 (1935).

[48] R.B. Griffiths, Phys. Rev. Lett. **24**, 715 (1970).

[49] A.N. Zisman, V.N. Kachinsky, V.A. Lyakhovitskaya, S.M. Stishov, ZhETF, **77**, 640 (1979) (English translation: A.N. Zisman, V.N. Kachinskii, V.A. Lyakhovitskaya, S.M. Stishov, JETP **50**, 322 (1979)).

[50] P. Bastie, J. Bornarel, J. Lajzerowicz, M. Vallade, Phys. Rev. **B12**, 5112 (1975).

[51] I.V. Aleksandrov, A.N. Zisman, S.M. Stishov, Phys. Solid State, **24**, 719 (1982).

[52] A.N. Zisman, V.N. Kachinsky, S.M. Stishov, ZhETF Pis'ma, **31**, 172 (1980) (English translation: A.N. Zisman, V.N. Kachinsky, S.M. Stishov, JETP Letters, **31**, 158 (1980)).

[53] P.V. Wigman, A.I. Larkin, V.M. Filev, JETP, **68**, 1883 (1975).

[54] S.M. Stishov, V.A. Ivanov, V.N. Kachinskii, ZhETF Pis'ma, **24**, 329 (1976) (English translation: S.M. Stishov, V.A. Ivanov, V.N. Kachinskii, JETP Letters, **24**, 297 (1976)).

[55] V.N. Kachinsky, V.A. Ivanov, A.N. Zisman, S.M. Stishov, ZhETF, **75**, 545 (1978)(English translation: V.N. Kachinskii, V.A. Ivanov, A.N. Zisman, S.M. Stishov, JETP **48**, 273 (1978)).

[56] G. Stanley, "Phase transitions and critical phenomena" (Clarendon Press, Oxford, 1971).

[57] R.Balescu, "Equilibrium and nonequilibrium statistical mechanics" John Wiley and Sons, 1975).

[58] J.-L. Barrat and J.-P. Hansen, "Basic concepts for simple and complex liquids" Cambridge, University Press (2003).

[59] A.P. Levanyuk, ZhETF, **36**, 810 (1959) (English translation: A.P. Levanyuk, JETP **9**, 571 (1959)).

[60] V.L. Ginzburg, Phys. Solid State, **2**, 2034, 1960.

[61] Neil Overend, Mark A. Howson, and Ian D. Lawrie, Phys. Rev. Lett. **72**, 3238 (1994).

[62] Sh. Ma, "The modern theory of critical phenomena" (W.A. Benjamin, Inc.1976).

[63] D. Kim, B. L. Zink, F. Hellman, S. McCall, G. Cao, and J. E. Crow, Phys. Rev. **B67**, 100406 R (2003).

[64] S.K. Japaridze, T.Sh. Kvirikashvili, V.A. Melik-Shakhnazarov, I.I. Mirzoeva, I.N. Makarenko, S.M. Stishov, JETP Letters, **54**, 62 (1991).

[65] A. Haupt and J. Straub, Phys. Rev. **E59**, 1795 (1999).

[66] J. A. Lipa, J. A. Nissen, D. A. Stricker, D. R. Swanson, T. C. P. Chui, Phys. Rev. **B68**, 174518 (2003).

[67] A. Kornblit, G. Ahlers, E. Buehler, Phys. Rev. **B17**, 282 (1978).

[68] M.J. Buckingham and W.M. Fairbank in "Progress in Low Temperature Physics" vol.3, p.80 ed. C.J. Gorter (North-Holland, Amsterdam, 1961).

[69] M.I. Bagatsky, A.V. Voronel, V.G. Gusak, JETP, **45**, 728 (1962).

[70] Yu.R. Chashkin, V.A. Popov, V.G. Simkin, A.V. Voronel, JETP, **45**, 828 (1963).

[71] L.P. Kadanoff, In Critical Phenomena, Proceedings of the Int. School of Physics, Enrico Fermi p. 101, Course LI, ed. M.S. Green (New York, Academic Press, 1971).

[72] A.Z. Patashinsky, V.L. Pokrovskii, "Fluctuation theory of phase transitions" (Pergamon, Oxford, 1979).

Appendix

In this section, I considered it necessary, at least briefly, at the level of the thesaurus to acquaint the inexperienced reader with some important concepts that they can meet while continuing to be educated. Corresponding information can help in avoiding a kind of "cultural" shock, almost inevitable in the study of more specialized literature.

A.1. Ising model

In view of the particular importance of the Ising model in the theory of phase transitions, we present the main results of the analysis of this model in the framework of the Bragg–Williams approximation. We recall that the Ising model is a system of N interacting spins located at the sites of a certain crystal lattice. Spin directions correspond to only two possible positions "up" or "down". The order parameter η in such a system can be determined from the relations:

$$\frac{N_+}{N} = \frac{1}{2}(1+\eta)$$

and

$$\frac{N_-}{N} = \frac{1}{2}(1 - \eta),$$

where N_+ and N_- are the number of spins directed up and down, respectively. According to Bragg and Williams, the free energy of the system in the Ising model has the form:

$$F = E - TS = -\frac{1}{2}zNJ\eta^2 + NkT\left\{\frac{1}{2}(1+\eta)ln\frac{1}{2}(1+\eta) + \frac{1}{2}(1-\eta)ln\frac{1}{2}(1-\eta)\right\}.$$

Here the first term is the total energy in the approximation of the nearest-neighbor interaction ($J > 0$-exchange integral, z-coordinate number). The value in curly brackets is the entropy S, defined as

$$S = klnW,$$

where $W = N!/N_+!N_-!$ is the number of distinguishable spin configurations (it is assumed that positive and negative spins form an ideal mixture). The phase transition of a ferromagnet-paramagnet in the Bragg–Williams approximation occurs at $T_c = zJ/k$ with a jump in the specific heat $\Delta c = 3/2zNJ/T_c = 3/2Nk$. As can be seen, the use of the Bragg–Williams method leads to the same result when studying phase transitions in the cases of two different systems: the Ising model and the binary alloy model (see Chapter 5).

A.2. Broken symmetry and Goldstone modes

The idea of broken symmetry actually goes back to the concept of L. Landau where the order parameter characterizes the spontaneous change in symmetry that occurs during the phase transition. Indeed, the physical system in the ground state, as a rule, has less symmetry than the symmetry of the microscopic

Hamiltonian describing the system as a whole. For example, the Hamiltonian of an isotropic ferromagnet is invariant with respect to three-dimensional rotations. However, in the ferromagnetic state, when all spins are aligned in one direction, the system becomes invariant only in a plane perpendicular to the magnetic moment. A phenomenon in which the ground state of a system does not have the symmetry of a Hamiltonian is called a "spontaneous symmetry breaking". In the case of continuous symmetry, the ground state of the system is degenerate, that is, there are an infinite number of ground states with the same energy (for example, states with different directions of the magnetic vector), but the system chooses one of them. As a result, "restoring symmetry" excitations of a very large wavelength with an energy tending to zero, arise at the wavelength tending to infinity. These excitations are called Goldstone modes. Known examples of the Goldstone modes include phonons in crystals and spin waves in ferromagnets, reflecting broken translational invariance and invariance with respect to rotations.

Additional information can be obtained from the books: Kerson Huang, Statistical Mechanics, John Wiley and Sons, 1987 (available online), R. Blints, B. Žekš, Soft Modes in Ferroelectrics and Anti-Ferroelectrics, North-Holland, 1974)

A.3. Soft modes

In structural phase transitions, a situation often arises where the frequency ω of any mode (branch) of the phonon spectrum tends to zero at the phase transition point for a certain value of the wave vector $q = q_0$. This corresponds to the freezing of the displacements determined by the vector q_0 and the appearance of a new crystal structure. This particular mode is called soft. The concept of a soft mode arose in connection with the development of the theory of ferroelectric phase transitions, but later it was also extended to other types of phase transformations (see R. Blints, B. Žekš, Soft Modes in Ferroelectrics and Anti-Ferroelectrics, North-Holland, 1974)

A.4. Kadanoff transforms

The block transformations of Kadanoff (Leo P. Kadanoff, 1966), which will be discussed below, played an important role in the development of the theory of phase transitions. In particular, they paved the way for the use of the renormalization-group approach to explain the similarity and universality relations and calculations of critical exponents.

Thus, consider a d-dimensional lattice with period a containing Ising spins with $S = \pm 1$ at the lattice sites. The Hamiltonian of the system in a homogeneous magnetic field directed along the z axis has the form:

$$H = -J_0 \sum_{\langle ij \rangle} S_i S_j - B \sum_i S_i, \tag{A.1}$$

where the quantities J_0 and B are the interaction constants. Next, we prepare spin blocks with the volume $(ba)^d$, where b is the scaling parameter. This procedure of "coarse graining", called the Kadanoff transformation, is illustrated in Fig. A.1. Denote the spin of each block as S_I, where I is the index of each block, and then

$$S_I = \frac{1}{N} \sum_{i \in I} S_i. \tag{A.2}$$

Here the sum of the spins of each block is normalized with the help of the parameter N, so that the values of the spin blocks correspond to the values ± 1. Simplifying the problem, Kadanoff attributes the value of the spin of the spin block $S_I = +1$, if the majority of spins in the block are looking up, and $S_I = -1$, respectively, in the opposite case. The justification of the "coarsening" procedure can be found in

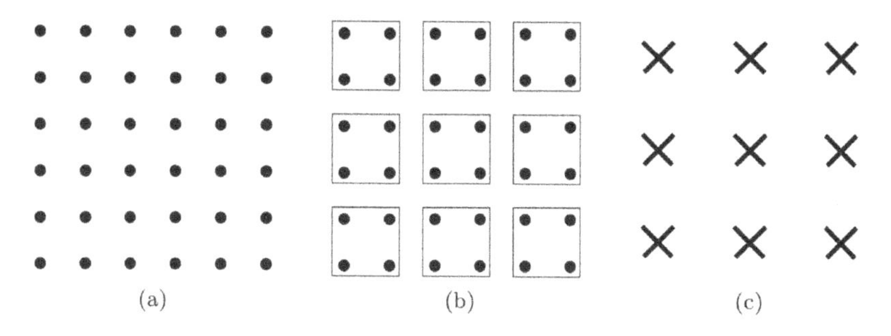

Figure A.1: Visualization of the Kadanoff block transformation. The original lattice of spins (a), divided into blocks consisting of 4 spins (b), is replaced by a lattice of effective spins (c) (see K. Wilson and J. Kogut, The renormalization group and the ϵ-expansion, Physics Reports, 12, 75, 1974).

the behavior of the correlation length ξ as we approach the critical temperature T_c. Indeed $\xi \to \infty$ for $T \to T_c$, so that we can assume that almost all the spins in the block are directed in one direction under the condition $ba \ll \xi$. The block Hamiltonian has the form:

$$H = -J_b \sum_{\langle IJ \rangle} S_I S_J - B_b \sum_I S_I, \tag{A.3}$$

where J_b and B_b are the new scaled-up interaction constants. The index b means that the system variables are converted using the scale factor b. As a result of this transformation, the number of spin blocks is represented by $N_b = b^{-d} N$, where N is the number of spin blocks before the transformation. Thus, the singular part of the free energy per block f_s near T_c is:

$$f_s(t,h) = b^{-d} f_s(t_b, h_b). \tag{A.4}$$

Hereinafter, the following notation is used: $h = H - H_c$; $m = M - M_c$; $t = (T - T_c)/T_c$. For further analysis, following Kadanoff, we introduce the following assumptions: $h_b = hb^x, x > 0$ $t_b = tb^y, y > 0$ and rewrite (A.4) in the form:

$$f_s(t,h) = b^{-d} f_s(tb^y, hb^x). \tag{A.5}$$

Form (A.5) means that the free energy per unit is a generalized homogeneous function, which indicates the scale invariance of the system. The values of x and y can not be calculated within the framework of the Kadanoff approach, but with their help, known similarity relations can be established. Differentiating (A.5), we have:

$$m(t,h) = b^{x-d}m(tb^y, hb^x), \tag{A.6}$$

$$\chi(t,h=0) = b^{2x-d}\chi(tb^y, hb^x), \tag{A.7}$$

$$C(t,h=0) = b^{2y-d}C(tb^y, hb^x). \tag{A.8}$$

For the case $h = 0, b = |t|^{-1/y}$, we obtain:

$$m(t,0) = (-t)^{-(x-d)/y}m(-1,0), \tag{A.9}$$

$$\chi(t,0) = (t)^{-(2x-d)/y}\chi(\pm 1.0), \tag{A.10}$$

$$C = (t)^{-(2y-d)/y}C(\pm 1). \tag{A.11}$$

It follows from (A.9–A.11):

$$\alpha = (2y-d)/y; \beta = (x-d)/y; \gamma = (2x-d)/y. \tag{A.12}$$

From (A.12) we obtain one of the similarity relations (see Chapter 5):

$$\alpha + 2\beta + \gamma = 0.$$

A.5. Renormalization group: basic concepts

Let us consider the Ising model, although the present analysis is more general. Near the critical point, the number of interacting degrees of freedom changes as $\sim \xi^d$, where ξ is the correlation length, d the space dimension. For $T \to T_c$, $\xi \to \infty$ and therefore, the number of degrees of freedom grows to infinity. In order to make the physical problem of the phase transition solvable it is necessary to reduce the number of degrees of freedom, say, from N to N' by integrating fluctuations with a wavelength λ corresponding to the inequality, $a \leq \lambda \leq ba$, where $b > 1$ is the spatial scale factor: $b^d = N/N'$. This procedure, called renormalization, can be carried out with the aid of block transformations of Kadanoff. In this case, naturally, the parameters of the Hamiltonian (coupling constants), such as $K = J/kT$ and $h = H/kT$ in the Ising model, change. The evolution of the parameter K (or the parameters $K1$, $K2$, etc. in a more complex case) as a result of successive iterations is usually illustrated using a current diagram.

Figure A.2 presents such a diagram for the case of the one-dimensional Ising model with the nearest-neighbor interaction. The diagram identifies the so-called fixed points, where the renormalization does not change the interaction parameters $K = K^*$, which means scale invariance. As is known, in this model there is no phase transition and the ordered state exists only at the point $T = 0$, which is an "unstable" fixed point. The latter means that the renormalization at a temperature slightly different from the zero value leads to a flow of the parameter K in the direction of the stable fixed point, as indicated by the arrows. The state of the system at this point for $K^* = 0$ and $T = \infty$ corresponds to complete disorder.

A different situation arises in the two-dimensional Ising model (see Fig. A.3). Here one more fixed point appears, indicating the existence of a phase transition (critical point) for K_c^*. This point is unstable, since renormalization at any temperature other than T_c leads to a flow in the direction of stable fixed points with $K^* = \infty$ and $K^* = 0$. A fixed point with $K^* = \infty$ corresponding to the full order is stable in this case. At the fixed point for K_c^*, the following

Figure A.2: Renorm-group picture of flow for the one-dimensional Ising model.

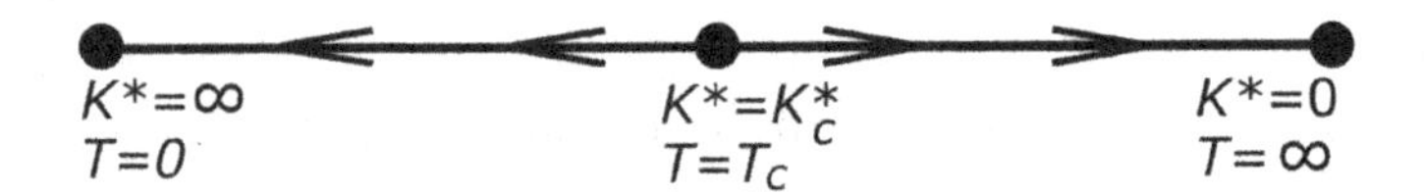

Figure A.3: Renorm-group picture of flow for two-dimensional Ising model.

equality holds:

$$\xi(K^*) = \xi(K^*)/b, \tag{A.13}$$

where ξ is the correlation length and b is the scale factor corresponding to the renormalization procedure (see Section 1.4). Equation (A.13) means that at a fixed point, $\xi(K^*) = 0$, or ∞. Accordingly, a fixed point with $\xi = \infty$ is called a "critical fixed point", while a fixed point with $\xi = 0$ is called "trivial". The critical exponents are determined by the specificity of the flow in the vicinity of the critical fixed point.

A few words about the so-called ε expansion and space in 3.99 dimensions. As is known, the critical exponents of the phase transition in the 4-dimensional space are practically equal to the mean-field values, which follow, for example, from the Landau theory (see Chapter 5).

Within the framework of the renormalization group approach, it was possible to express the values of the critical exponents in the form of an expansion in powers of small $\varepsilon = 4 - d$. The dimension of the space d is assumed to be equal to the fractional value 3.99. As a result, for example, the corresponding expansion for the exponent α, which determines the behavior of the heat capacity in the critical region, has the form:

$$\alpha = -\frac{4-n}{2(n+8)}\varepsilon + \frac{(n+2)^2(n+28)}{4(n+8)^2}\varepsilon^2.$$

Here, n is the dimension of order parameter. The most surprising thing about this expansion is that for $n = 3$, and substituting $\varepsilon = 1$, we obtain the value $\alpha = -0.1$, which agrees with the experimental value (see Table 2, Chapter 5). The detailed information in relation of paragraphs A.4 and A.5 can be found in the following references:

J. M. Yeomans, Statistical Mechanics of Phase Transitions, Oxford University Press, 1992 (available on the Internet).

Nigel Goldenfeld, Lectures on Phase Transitions and the Renormalization Group, Perseus Books, Reading, Massachusetts, 1992 (available on the Internet).

John Cardy, Scaling and Renormalization in Statistical Physics, Cambridge University Press, 2002 (available on the Internet).

Sh.Ma, Modern Theory of Critical Phenomena, W.A. Benjamin, Inc. (1976).

A.6. Fluctuations and phase transitions

S.M. Stishov

Institute for High Pressure Physics of Russian Academy of Sciences, Troitsk, Moscow, 2005

The role of classical and quantum fluctuations at the second order phase transitions is discussed. It is asserted that the definition of classical and quantum phase transitions as transitions that occur as a result of thermal and, respectively, quantum fluctuations is quite legitimate.

This note is written under the influence of a number of discussions on the role of fluctuations in phase transitions, which occurred in connection with the publication of an article on quantum phase transitions [1]. As is known, the behavior of physical quantities at the second-order phase transition is strongly influenced by intensive fluctuations of the order parameter. This role of fluctuations is widely covered in the educational and monographic literature (see, for example, [2, 4, 5]). However, it turns out that there is some ambiguity about a more general question related to the global role of fluctuations at the phase transitions. When looking at the relevant literature, it may appear that fluctuations are a kind of decoration not having any relation to the phase transition mechanism. On the other hand, one can often hear and read statements that link the onset of classical and quantum phase transitions, respectively, with thermal and quantum fluctuations [6–10].[1,2] These statements often caused objections, some of which undoubtedly have a semantic basis. Nevertheless, it seems useful, at least from a didactic point of view, to try to answer more clearly the questions about why phase transitions occur in general, or in other words, the reason for vanishing the order parameter at some critical value of the intensive or field variable (T, P, H, etc.) and, accordingly, the role of the fluctuations.

In conclusion of the introductory part, we note that when considering the role of fluctuations at phase transitions, it is obvious that its kinetic and thermodynamic aspects should be distinguished. As for kinetics, as is well known, phase transitions of first order do not occur until the corresponding nucleus is formed as a result of fluctuations, thermal or quantum. In the case of phase transitions of second order, the situation is less obvious. However, it should be remembered that any system finds its energy minimum only due to fluctuations, and therefore fluctuations, irrespective of their thermodynamic contribution, play a crucial role in the evolution of the corresponding system.

---

[1]As an example, we quote from the book [9]: We note that phase transitions in classical models are driven only by thermal fluctuations, as classical systems usually freeze into a fluctuationless ground state at $T = 0$. In contrast, quantum systems have fluctuations driven by the Heisenberg uncertainty principle even in the ground state, and these can drive interesting phase transitions at $T = 0$.

[2]The absence of long-range order in a number of low-dimensional systems with strong developed fluctuations, obviously, support these statements.

Classical phase transitions

Phase transitions occurring at $T > 0$ can always be described within the framework of classical statistical mechanics.[3] This also applies to such essentially quantum phenomena as superfluidity and superconductivity. In this connection, any phase transitions occurring at $T > 0$ are called classical phase transitions [1, 7, 8]. So, let us consider the most usual case of a phase transition occurring upon variation of temperature T. The thermodynamic potential of the system:

$$G = H - TS = E + PV - TS \tag{A.14}$$

determines the entropy as the thermodynamically conjugate quantity for temperature. Apparently, entropy should be considered a characteristic parameter of our problem. In the context of the present article, it is important to note that all quantities, except for entropy, entering into the relation (A.14) are either external parameters, or average over the Gibbs ensemble. Entropy, being a logarithm of the phase volume, characterizes the statistical ensemble as a whole. Accordingly, the phase volume of the ensemble includes regions accessible only as a result of fluctuations. The Landau theory establishes a connection between the entropy S and the order parameter η in the form [2]:

$$S = -\left(\frac{\partial G}{\partial T}\right)_P = S_0 - \frac{\partial A}{\partial T}\eta^2 \tag{A.15}$$

where $A(P,T) = a(P)(T - T_c)$. Usually $a > 0$, which corresponds to the case when the symmetric or disordered phase has a greater

---

[3]In the nearest neighborhood T_c, there is a region, where the inequality $h\omega^* < kT$ holds, therefore corresponding to the classical behavior of critical fluctuations, here ω^* is the characteristic frequency of fluctuations tending to zero, when approaching T_s.

entropy and is located on the temperature scale for $T > T_c$. If the phase transition occurs under isothermal conditions with a changè of pressure, the characteristic parameter of the problem is the volume V as the value conjugate to the pressure P.[4] However, the volume variation is led only to a renormalization of energy and entropy, which in any case does not change the nature of the phase transition as a result of the competition between energy and entropy. Further, in order to understand the physical meaning of the quantities under discussion and the relation between entropy and fluctuations, it is useful to turn to the analysis of the microscopic model of the phase transition. We use for this purpose the Ising model, which is a system of N interacting spins located at the sites of a certain crystal lattice. In the absence of a magnetic field, the Ising model is described by the Hamiltonian:

$$H = -\sum_{i,j}^{N} J_{ij}\hat{\sigma}_i^z\hat{\sigma}_j^z \tag{A.16}$$

where J_{ij} is the exchange integral, $\hat{\sigma}_i, \hat{\sigma}_j$ are the spin operators. The eigenvalues of the operator $\hat{\sigma}^z$ take values ± 1 corresponding to the two possible directions of the spin "up" and "down". We emphasize that the Ising model in the absence of a magnetic field or in a field parallel to the z axis is classical in view of the commutativity of the operators $\hat{\sigma}^z$. The order parameter η in such a system can be determined from the relations:

$$\frac{N_+}{N} = \frac{1}{2}(1+\eta); \frac{N_-}{N} = \frac{1}{2}(1-\eta), \tag{A.17}$$

where N_+ and N_- is the number of spins directed up and down. For our purpose, it suffices to use the results of the study of the Ising

[4]In the framework of the Landau theory, the volume can be written in the form $V = -(\frac{\partial F}{\partial P})_T = V_0 - \frac{\partial A}{\partial P}\eta^2$, which indicates the connection of the excess volume $V - V_0$ and of excess entropy $S - S_0$.

model within the framework of the mean-field concept in the Bragg–Williams approximation. In the presentation of the results of this approximation, we will follow R. Kubo [3]. According to Bragg and Williams, the free energy of the system in the Ising model has the form:

$$F = -\frac{1}{2}zNJ\eta^2 + NkT\left\{\frac{1}{2}(1+\eta)ln\frac{1}{2}(1+\eta) + \frac{1}{2}(1-\eta)ln\frac{1}{2}(1-\eta)\right\}. \quad (A.18)$$

Here, the first term is the total energy in the approximation of the nearest-neighbor interaction ($J > 0$-exchange integral, z-coordinate number). The value in curly brackets is the entropy S, defined as

$$S = klnW, \quad (A.19)$$

where

$$W = \frac{N!}{N_+!N_-!}, \quad (A.20)$$

W is the number of distinguishable spin configurations (it is assumed that positive and negative spins form an ideal mixture). The equilibrium value of the order parameter η for each given temperature is determined from the condition $\partial F/\partial\eta = 0$. The phase transition of a ferromagnet-paramagnet in the Bragg–Williams approximation occurs at $T_c = zJ/k$.[5]

Note that although the order parameter η and the entropy S are expressed in terms of the same variables (see relations A.17–A.20), there is a fundamental difference between these quantities. As can be seen from (A.17–A.20), to determine the order parameter it is sufficient to know only the total number of spins in the "up"

[5]Note that the exact solution for the two-dimensional Ising model with the nearest-neighbor interaction in the case of a square lattice leads to a critical temperature $T_c = 0.57zJ/k$ [5].

or "down" positions, then to calculate the entropy it is necessary to take into account all the spin configurations corresponding to a given order parameter. This situation is absolutely not related to the specifics of a particular model, but reflects the essence of physical systems in general and the statistical nature of entropy in particular. So, within the framework of the mean-field approach, even in a state of equilibrium, the spin configuration of the system continuously changes, leaving the order parameter unchanged.[6] The question is, can we call this "flicker" of configurations a fluctuation? At first glance, no, because we call fluctuation deviations from the mean values. In our case, all configurations are equal and the concept of the average configuration does not make sense. However, it is not difficult to guess that transitions between different spin configurations arise as a result of fluctuation rotations (flips) of one or several spins, which means the existence of local density fluctuations of differently directed spins. In other words, if you try to track the number of spins directed up and down in a certain volume small in comparison with the total volume of the system, then we see that these numbers fluctuate, thus causing transitions between different spin configurations and hence the subsequent local fluctuations of the order parameter.

We note further that in mean-field approximations only the pair fluctuation rotations of the spins are taken into account, leaving the order parameter unchanged. In higher approximations, all possible rotations are taken into account, including those leading to fluctuations of the order parameter, however, as we have seen above, local fluctuations of the order parameter occur even in the case of "paired" fluctuations.[7]

---

[6] In the real case, and far from the critical point, the phrase "almost unchanged" should be used.

[7] Obviously this is one of the differences between the feminological Landau theory and approximations based on the mean field concept.

Quantum phase transitions

A classical system, cooled to 0 K, passes into a state with no fluctuations at all, in which nothing happens because of the absence of kinetic energy. In the quantum case, the situation is completely different. The energy of the ground state of a bound or condensed system of particles in the general case contains a fluctuation part, often called "zero point" energy. The origin of zero energy is related to the non-commutativity of the corresponding operators and the Heisenberg uncertainty relations that follow from this. Accordingly, the free energy of the system at $T = 0$ can be conditionally written in the form[8]

$$F = E_{st} + E_z, \tag{A.21}$$

where E_{st} and E_z are the static and zero energy of the system, respectively. The ratio E_{st}/E_z in the general case does not remain constant with the variation of external parameters, which determines its evolution with a change in pressure, magnetic field, chemical composition, etc. The role of zero point energy E_z is to some extent analogous to the role of the entropy term in (1). Indeed, the occurrence of spatial disorder (or, more accurately, delocalization of particles) contributes to a decrease in the zero energy of the system, which is in fact the factor providing a phase transition. Turning again to the Ising model, we recall that the latter acquires quantum features in a transverse magnetic field. The corresponding Hamiltonian containing two non-commuting operators $\hat{\sigma}_i^z$ and $\hat{\sigma}_j^x$ has the form:

$$H = -\sum_{i,j}^{N} J_{ij}\hat{\sigma}_i^z\hat{\sigma}_j^z - \Gamma\sum_{i}^{N}\hat{\sigma}_i^x, \tag{A.22}$$

where Γ is the transverse field. The commutator $[H\hat{\sigma}_i^z]$ has finite value for $\Gamma \neq 0$. This circumstance causes the existence of zero (quantum)

[8]Such an entry is meaningful only in cases when quantum effects are small.

fluctuations in the spin direction, the intensity of which increases with increasing Γ and ultimately leads to a phase transition at the critical value $\Gamma = \Gamma_c$ (see, for example, [11–13] and the experimental work [14]). The Heisenberg Hamiltonian:

$$H = -\sum_{i,j}^{N} J_{ij}\hat{\sigma}_i\hat{\sigma}_j, \tag{A.23}$$

where $\hat{\sigma}_i\hat{\sigma}_j = \hat{\sigma}_i^x\hat{\sigma}_j^x + \hat{\sigma}_i^y\hat{\sigma}_j^y + \hat{\sigma}_i^z\hat{\sigma}_j^z$, initially contains non-commuting operators, which basically ensures the existence of quantum fluctuations in the corresponding models with localized spins. In particular, the ground state of the Heisenberg antiferromagnet contains a fluctuation term of the form $\sum \frac{1}{2}h\omega_q$ corresponding to the zero point magnon energy (ω_q is the spin wave frequency).[9] For this reason, under certain conditions, the system can transform into a paramagnetic state through a quantum phase transition (see, for example, [15]). It seems that everything stated above positively solves the question of the legitimacy of determining the classical and quantum phase transitions as transitions that occur as a result of thermal and respectively quantum fluctuations.

References

[1] S.M. Stishov, UFN, **174**, 853 (2004).
[2] L.D. Landau, E.M. Lifshitz, Statistical physics, part 1 (Pergamon Press, Oxford 1980).
[3] R. Kubo, Statistical mechanics, translation from English (North Holland, 1965).
[4] A.Z. Patashinskii, V.L. Pokrovskii, Fluctuation theory of phase transitions (Pergamon Press, Oxford, 1979).

[9]In the ground state of a Heisenberg ferromagnet, there are no fluctuation terms. For this reason, no events can occur in an isolated ferromagnetic spin system at $T = 0$. In real systems, the situation can change with the interaction of spins with other degrees of freedom.

[5] G. Stanley, Phase Transitions and Critical Phenomena (Clarendon Press, Oxford, 1971).
[6] Nigel Goldenfeld, Lectures on Phase Transitions and the Renormalization Group (Perseus Books, Reading, Massachusetts, 1992).
[7] Cardy John, Scaling and Renormalization in Statistical Physics (Cambridge University press, 2002).
[8] Sondhi S.L., Girin S.M., Carini J.P., Shahar D., Rev. Mod. Phys., **69**, 315 (1997).
[9] Sachdev S., Quantum Phase Transitions, (Cambridge University Press, UK, 1999).
[10] Continentino M.A., Quantum Scaling in Many-Body Systems (World Scientific, Singapore, 2001).
[11] Vojta Thomas, Ann. Phys.(Leipzig), **9**, 443 (2000).
[12] Pfeuty P. and Elliott R.J., J. Phys. C: Solid St. Phys., **4**, 2370 (1971).
[13] Young A.P., J. Phys. C: Solid St. Phys., **8**, L309 (1975).
[14] Bitko D., Rosenbaum T.F., Aeppi G., Phys. Rev. Lett., **77**, 940 (1996).
[15] Chakravarty S., Harperin B.I., and Nelson D.R., Phys. Rev., B **39** 2344 (1989).

A.7. Quantum phase transitions

The next section introduces the reader to the so-called quantum phase transitions and quantum criticalities. As educational material for the presentation of this subject is my fairly popular article published some time ago in the journal "Uspekhi Fizicheskikh Nauk." Here I would like to correct one historical mistake. The point is that J. Hertz was not the first investigator of the problem of phase transitions occurring at $T = 0$, as indicated in the article. In fact, V. Vaks and M. Geilikman (JETP, 60, 330, 1971) and A. Rechester (JETP, 60, 782, 1971) considered this problem long before Hertz. In addition, I have also enclosed two more review articles, which are devoted to hot topics in the subject.

Physics – Uspekhi **47** (8) 789 – 795 (2004)

REVIEWS OF TOPICAL PROBLEMS

PACS numbers: **64.60.–i, 71.10.–w**, 71.10.Hf

Quantum phase transitions

S M Stishov

DOI: 10.1070/PU2004v047n08ABEH001850

Contents

Abstract. Continuous phase transitions that occur at zero temperature as a result of quantum fluctuations required by Heisenberg's uncertainty principle are called quantum phase transitions. In the present paper an elementary introduction to quantum phase transitions is given. A few experimental examples from the physics of heavy-fermion systems and itinerant ferromagnets are described.

1. Introduction

Recent studies in the field of strongly correlated electron systems have largely been concentrated on the so-called quantum phase transitions or quantum critical phenomena. And although a number of monographs and numerous reviews [1 – 9] have already been devoted to this comparatively new problem, it is still very far from being completely resolved.

As distinct from classical phase transitions, quantum phase transitions result from nonthermal quantum fluctuations, which arise due to the uncertainty principle [1]. The concept of quantum phase transitions was first introduced by J Hertz in 1976 [10] [2], who showed that in view of the inextricable relation between the static and dynamic properties of a quantum system its time characteristics significantly affect the behavior of the substance in the critical region at $T = 0$ and the effective dimension of a quantum system always exceeds its spatial dimension [3]. The latter fact largely influences the behavior of a substance in the critical region.

Quantum phase transitions in a pure form only occur at $T = 0$, although their effect on the properties of the substance can spread to the region of finite temperatures. We will emphasize that here and in what follows we only deal with continuous or second-order phase transitions. First-order quantum phase transitions (e.g., helium melting) do not have a fluctuation region and are of no interest in this context.

Further, we will need a brief review of classical phase transitions, which will be presented in the next section.

2. Brief review of phase transitions

The theory of second-order phase transitions developed by Landau is based on the concept of the order parameter η (for more details concerning second-order phase transitions see [14 – 16]). In the framework of the Landau theory the order parameter η near the transition point behaves as

$$\eta \propto |t|^{1/2}, \tag{1}$$

where $t = (T - T_c)/T_c$.

Heat capacity, the coefficient of thermal expansion, and compressibility experience finite jumps at the transition point.

In the general case, t in (1) should be replaced by the quantity δ which is a certain dimensionless distance to the phase transition point and is expressed, depending on the particular case, in terms of pressure, magnetic field strength, concentration, etc. The behavior of the thermodynamic (physical) quantities $y_1, \ldots, y_N$ in the vicinity of a phase transition point is typically expressed by power functions of the form

$$y \propto |\delta|^{-x}, \tag{2}$$

[1] Although generally accepted, this formulation should obviously be explained. It means that quantum fluctuations, which destroy the long-range order in a system, are controlled by non-thermal parameters like pressure, concentration, magnetic field etc.

[2] Part of the results of that paper were revised in Ref. [11].

[3] This circumstance was probably first noted in Ref. [12] (see also [13]) in the analysis of the quantum two-dimensional Ising model with a spin 1/2 in a transverse magnetic field.

S M Stishov Institute for High Pressure Physics,
Russian Academy of Sciences,
142190 Troitsk, Moscow Region, Russian Federation
Tel. (7-095) 334-00 10
Fax (7-095) 334-00 12
E-mail: sergei@hppi.troitsk.ru, stish@ips.ras.ru

Received 15 March 2004, revised 30 April 2004
Uspekhi Fizicheskikh Nauk **174** (8) 853 – 860 (2004)
Translated by M V Tsaplina; edited by M V Chekhova

where x is the so-called critical exponent determined from the relation

$$x \equiv \lim_{\delta \to 0} \frac{\ln y(\delta)}{\ln \delta}\,. \tag{3}$$

With such a definition, the critical indices of 'nonsingular' quantities, quantities with logarithmic divergence, or quantities with a 'peak-like' singularity turn out to be equal to zero.

In a simple version, the Landau theory totally ignores spatial fluctuations of the order parameter and is therefore invalid in a region close to a phase transition. However, in some cases the fluctuation region is so narrow that it cannot be resolved in a real experiment. The simplest way to allow for fluctuations is to expand the thermodynamic potential, not only in the power series of the order parameter, but also in its gradients. The first significant term of the gradient expansion is of the form

$$\delta \Phi \propto (\Delta \eta)^2 \simeq \xi^2 \left(\frac{\partial \eta}{\partial x}\right)^2. \tag{4}$$

The quantity ξ with dimensionality of length is called the correlation length and characterizes the spatial inhomogeneity of the system. In the framework of the Landau theory the correlation length has the form

$$\xi = \xi_0 |t|^{-1/2}. \tag{5}$$

In the general case the correlation length is written as

$$\xi = \xi_0 |t|^{-\nu}. \tag{6}$$

Using relation (4) one can estimate the width of the fluctuation region and, correspondingly, the range of applicability of the Landau theory from the expression for the relative mean-square order parameter fluctuation (see [17, 18]),

$$\frac{(\Delta \eta)^2}{\eta^2} = \frac{T_c^2}{\Delta C_p \xi_0^d (T - T_c)^{2-d/2}}, \tag{7}$$

where d is space dimensionality.

Expression (7) also establishes an important relation between the space dimensionality and the intensity of fluctuations. From (7) it follows that for $d \geqslant 4$ the order parameter fluctuations are finite for any t and, accordingly, diverge for $d < 4$. The dimension $d = 4$ is in this case called the upper critical dimension — d_c^+. A lower critical dimension d_c^- also exists, for which the long-range order is absent in the system at any finite temperature.

Note that at the tricritical point

$$\frac{(\Delta \eta)^2}{\eta^2} \propto \frac{T_c}{\xi_0^d (T - T_c)^{(3-d)/2}}, \tag{8}$$

the upper critical dimension is $d_c^+ = 3$.

Let us now focus our attention on the relation between static and dynamic phenomena in the critical region:

$$\tau \propto \xi^z, \tag{9}$$

where τ is the relaxation time of the order parameter and z is the dynamic critical exponent.

It is appropriate to denote τ as ξ_τ and rewrite (9) in the form

$$\xi_\tau \propto \xi^z. \tag{10}$$

Relations (6) and (10) imply

$$\xi_\tau \propto |t|^{-z\nu} \tag{11}$$

or

$$\omega^* \propto |t|^{z\nu}, \tag{12}$$

where ω^* is the characteristic frequency of fluctuations.

3. What are quantum phase transitions?

Phase transitions occurring at $T = 0$ upon variation of variables determining the intensity of quantum fluctuations are called quantum phase transitions.

We will consider Fig. 1, where $T_c(P)$ is the line of a continuous phase transition corresponding, for example, to a certain magnetic transformation. At a normal pressure, the phase transition temperature has an entirely finite value. An increase in pressure leads to a progressive fall in the transition temperature, down to $T = 0$ K at a certain critical pressure P_c (another control parameter, for instance, the magnetic field, concentration, etc., may be used here instead of pressure). The 'disordered' phase due to quantum fluctuations at $T = 0$ is a realization of 'quantum' disorder, which is essentially different from 'classical' disorder. In particular, for a magnetic quantum phase transition, the paramagnetic phase cannot be treated as a system of individual spins fluctuating, for example, between 'up' and 'down' states in real time. The ground state of a quantum paramagnet is described by the wave function, which is a quantum superposition of these states and therefore possesses zero entropy [1].

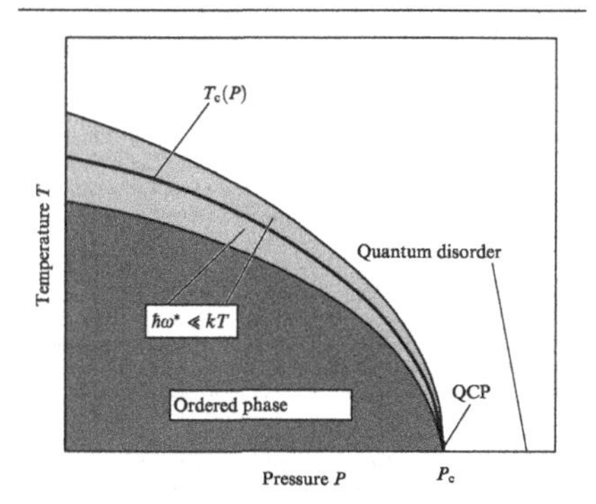

Figure 1. Schematic phase diagram of a substance undergoing a second-order phase transition where there is a negative slope of the transition curve $dT/dP < 0$. ω^* is the characteristic frequency of fluctuations, which vanishes at the phase transition point. The region around the phase transition line $T_c(P)$ is the region of classical fluctuations. $dT/dP = \infty$ at the quantum critical point (QCP) for $P = P_c$.

In this case, the quantum phase transition at $T = 0$ is a limiting case of the classical phase transition that occurs at $T \neq 0$. However, as has been mentioned above, the situation is possible (a system in a state of the lower critical dimension) where a phase transition may occur at $T = 0$ [4] only.

Phase transitions occurring at $T > 0$ can always be described in the framework of classical statistical mechanics (henceforth, the description of a phase transition is understood as an analysis of the behavior of the order parameter, correlation functions, and thermodynamic quantities in the vicinity of T_c). This also applies to such essentially quantum phenomena as superfluidity and superconductivity. The cause can readily be understood with the help of relation (12), implying that as $t \to 0$, that is, in the nearest neighborhood of T_c, the following inequality always holds:

$$\hbar\omega^* \ll kT, \tag{13}$$

which corresponds to the classical behavior of critical fluctuations. According to (13) the selected region around the phase transition line is the region of classical fluctuations [19, 20]. This naturally does not mean that quantum mechanics plays no role in this case. Quantum mechanics determines the very existence of the order parameter, whereas its behavior in the critical region [5] at $T > 0$ is controlled precisely by the classical thermal fluctuations.

It is a known fact [1, 4, 10, 12, 13, 19, 20] that the quantum statistical problem of a phase transition in a d-dimensional space at $T = 0$ can be reduced to the classical problem [6] with effective dimension $(d+1)$. Imaginary time in the interval $[0, -i\hbar\beta]$, where $\beta = 1/kT$, stands as an additional coordinate here. Generally, the coordinate space of such a system is finite in the direction of time, but as $T \to 0$ the time interval becomes infinite and the system acquires all features of a classical system in a $(d+1)$- dimensional space. However, when it concerns the critical properties of the system, its effective dimension appears to be equal to $d + z$, where z is the dynamic exponent. So, the effective dimension of a quantum system in the critical region at $T = 0$ may appear to equal or even exceed the upper critical dimension d_c^+ with all ensuing consequences (see above). This effect is illustrated in Fig. 2, which shows the results of the analysis of the evolution of the exponent β corresponding to the critical behavior of the order parameter (magnetization in this case) upon the phase transition in the antiferromagnet $MnCl_2 \cdot 4H_2O$ [21]. One can see in Fig. 2 that, as the temperature drops, the exponent β increases and tends to the mean-field value $\beta = 0.5$.

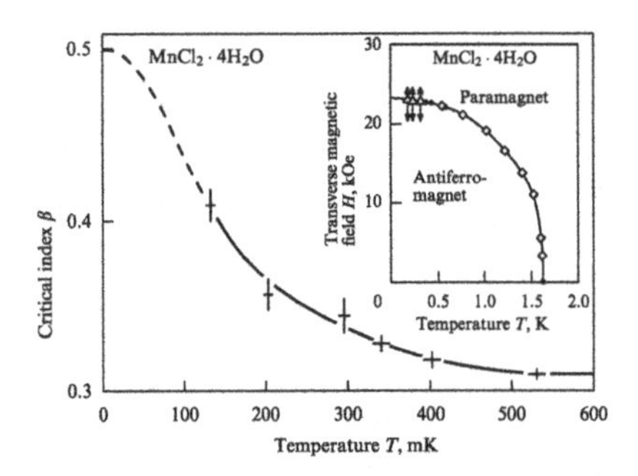

Figure 2. Dependence of the critical exponent β determining the behavior of the order parameter on the temperature of measurement for the antiferromagnet $MnCl_2 \cdot 4H_2O$ [21]. The inset gives the corresponding phase diagram in magnetic field H — temperature T coordinates.

One should also bear in mind that for $\xi_\tau < L_\tau$, where $L_\tau = \hbar/kT$ is the time extent of the space-time continuum, the system is unaware of being at a finite temperature and behaves as if it were in a $(d+1)$-dimensional space. The lines $\xi_\tau = L_\tau$ conditionally divide the phase diagram of the substance with a quantum critical point (QCP) into regions with different effective dimensions. The same lines mark the crossover between phenomena occurring at small and large times characterized by the correlation time ξ_τ (Fig. 3). We should stress that for $\xi_\tau < L_\tau$ the description of the phenom-

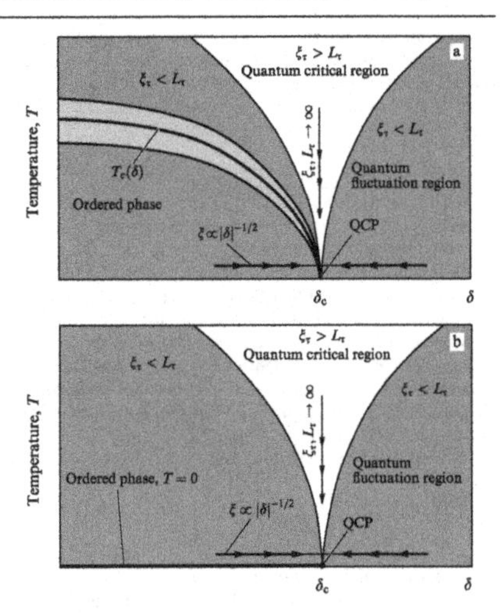

Figure 3. Schematic phase diagram of a substance in the vicinity of a quantum critical point: (a) ordered phase exists at finite temperatures; (b) ordered phase exists at $T = 0$ only. ξ is the correlation length, $\xi_\tau \propto \xi^z$ is the correlation time, $L_\tau = \hbar/kT$ is the extent of time coordinate. The lines $\xi_\tau = L_\tau$ correspond to the quantum-classical crossover.

[4] This is the case, for example, with the one-dimensional Ising model.

[5] Quantum fluctuations are undoubtedly very important at distances on the order of interatomic ones, but fluctuations of much larger scale, with a correlation length of dozens and hundreds of interatomic spacings, which control the behavior of the system in the critical region at $T > 0$, are described adequately within classical statistical mechanics [4, 5, 19, 20].

[6] Recall in this connection that in the framework of fluctuation theory the values of critical exponents characterizing a phase transition are independent of the microscopic nature of a substance, but are defined by the symmetry of the Hamiltonian and the dimensionality of space, which determines the particular class of universality [15, 16, 19]. It has turned out that the critical exponents of the Ising quantum model in a transverse magnetic field at $T = 0$ correspond to a higher space dimension than the initial dimension of the problem [12, 13]. It is precisely this situation that was explained in the pioneering paper by Hertz [10].

ena requires a quantum-mechanical approach, whereas for $\xi_\tau > L_\tau$ the corresponding phenomena become classical because of the loss of phase coherence.

Figure 3 presents schematically the phase diagrams of substances possessing QCP for which two cases are possible: the ordered phase exists at $T = 0$ only (Fig. 3a) or the ordered phase exists also at $T > 0$ (Fig. 3b). The lines corresponding to the condition $\xi_\tau = L_\tau$ separate regions with predominantly classical and predominantly quantum fluctuations.

Thus, in the phase diagrams one can distinguish between regions corresponding to ordered and quantum-disordered (quantum-fluctuation) states of a substance and a region of mixed nature, referred to as the quantum critical region. Since practically always $d + z \geqslant 4$, the behavior of the correlation function in the former two regions corresponds to the Gaussian case $\xi \propto |\delta|^{-1/2}$. In the mixed region, along the trajectory $\delta = 0$ the behavior of the correlation function is controlled exclusively by the temperature, which indicates the absence of any other energy scales. For metallic systems it means that the Fermi energy E_F (or the Fermi temperature T_F) no longer plays the role of a universal scaling factor in the description of the electronic properties of materials in a quantum critical region, and accordingly the absolute temperature assumes this role as the energy scale. As will be seen below, this situation is due to the divergence of the effective mass of carriers at the QCP. It is this fact that determines the so-called non-Fermi-liquid behavior [22–24]. In systems whose effective dimension is less than the upper critical dimension, $d < d_c^+$, the absolute temperature determines the scale of all phenomena in the quantum critical region. This situation is described in the English literature as E/T or ω/T scaling (see, e.g., [6, 7]).

Section 4 presents a description of experimental examples illustrating some of the above statements. However, we will first give necessary explanations concerning the thermodynamics of phase diagrams of substances with quantum critical behavior. Figure 1 schematically shows a particular case where the phase transition line $T_c(P)$ at $T > 0$ has a negative slope, $dT_c/dP < 0$. This type of behavior is rather widespread but not the only possible. Cases are possible when dT_c/dP reverses sign at a certain positive pressure or when the derivative dT_c/dP is always positive. The latter case is of no interest in the present context. Naturally, the aforesaid is also valid when a control parameter other than pressure is considered.

We shall note further that the slope of the phase transition curve tends to infinity with approaching the QCP[7] ($dT_c/dP \to \infty$ as $T_c \to 0$). Similarly, for the phase diagram in coordinates T-H (H is the magnetic field) one can write $dT_c/dH \to \infty$ as $T_c \to 0$. If the concentration e of the impurity element is taken as the 'control' parameter, the situation becomes somewhat more complicated, but when the impurity, creating the so-called 'chemical pressure', plays to an extent a passive role, we can again write $dT_c/dc \to \infty$ as $T_c \to 0$. It should also be recalled that, as follows from the Nernst heat theorem, all temperature derivatives of thermodynamic quantities are equal to zero at the absolute zero of temperature[8]. Accordingly, the amplitudes of anomalies (jumps) of specific heat, thermal expansion coefficients, etc. observed during a high-temperature phase transition tend to zero when approaching the QCP. Hence, the results of measurements of specific heat and other thermal quantities along a trajectory corresponding to the critical coordinate $\delta = \delta_c$ are not perturbed by the nearness of the phase transition line.

Concluding this section, we will emphasize that in the classical case the phase transition temperature does not become zero for a finite value of the parameter δ, and therefore the very shape of the phase transition curves shown in Figs 1 and 3 is a manifestation of quantum effects.

4. Examples of systems with quantum critical behavior

4.1 Heavy-fermion compounds[9] $CeCu_{6-x}Au_x$, $YbRh_2Si_2$, $YbRh_2(Si,Ge)_2$, $CePd_2Si_2$, and $CeIn_3$

(i) The compound $CeCu_{6-x}Au_x$ has an antiferromagnetic ground state for $x < 0.1$ with a Neel temperature T_N that rises linearly with increasing x up to $x = 1$ (Fig. 4) [26, 27]. In the neighborhood of the quantum critical point for $x = x_c \approx 0.1$ the behavior clearly differs from that of a Fermi liquid[10] [27–29]. In particular, in the quantum critical region the heat capacity varies as $C/T = a\ln(T_0/T)$ and the static magnetic susceptibility depends on the temperature

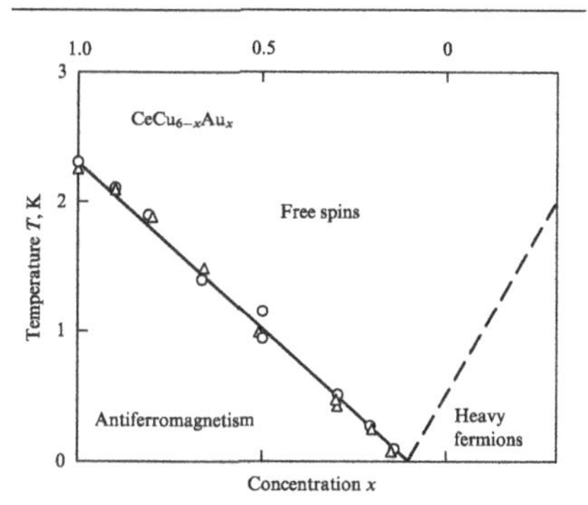

Figure 4. Concentration phase diagram of $CeCu_{6-x}Au_x$ [28]. Separate regions are named according to the nomenclature used in paper [28]. In terms of the diagrams presented in Fig. 3a, b, the free-spin region corresponds to the quantum-critical area.

[7] This conclusion is a corollary of the Nernst heat theorem stating that the entropy S is identically equal to zero at $T = 0$. From this, for first-order phase transitions we obtain directly from the Clausius–Clapeyron equation $dT_c/dP \to \Delta V/\Delta S$ (ΔV and ΔS are volume and entropy jumps upon the phase transition), $dT_c/dP \to \infty$ as $T_c \to 0$. For second-order phase transitions one should use one of the Ehrenfest equations: $dT_c/dP = \Delta(dV/dP)/\Delta(dS/dP)$. From the identity $S \equiv 0$ at $T = 0$ we again obtain $dT_c/dP \to \infty$ as $T_c \to 0$. However, the temperature at which the corresponding effects are observed can be fairly low.

[8] In this connection, the statement made in [25] about the divergence of the coefficient of thermal expansion at a quantum critical point looks rather strange.

[9] Heavy-fermion compounds constitute a class of metallic materials with a strong electron correlation that possess a chemically ordered lattice of magnetic ions (a Kondo lattice). In these compounds, conduction electrons interact with local magnetic moments of the ions. As a result, the effective mass of conduction electrons becomes very large [26].

[10] In the case of a normal Fermi liquid we have $C \sim \gamma T$ or $C/T = \text{const}$; $\chi = \text{const}$ and $\Delta\rho \sim T^2$.

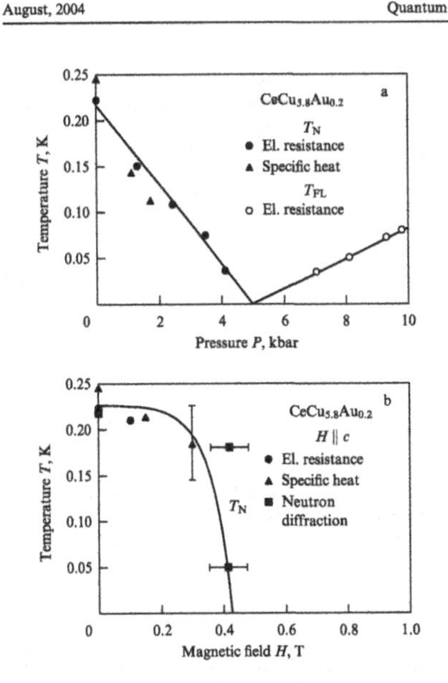

Figure 5. Phase diagram of the low-temperature antiferromagnet $CeCu_{5.5}Au_{0.2}$: (a) at high pressures; (b) in a magnetic field. T_{FL} marks the boundary of the Fermi-liquid regime [29].

according to the expression $\chi^{-1} = \chi_0^{-1} + cT^a$, $a < 1$; the electrical resistance behaves quasilinearly, namely, $\Delta\rho = \rho - \rho_0 \sim T^m$, $m \approx 1$. For the compound $CeCu_{6-x}Au_x$ the quantum critical point can be reached under hydrostatic pressure (e.g., for $x = 0.2$ the critical pressure is $P_c \approx 5$ kbar, Fig. 5a), and the corresponding temperature dependences have essentially the same form as dependences obtained in a 'chemical' realization of the QCP. The situation is different when the quantum critical point in $CeCu_{6-x}Au_x$ is attained through a variation of the magnetic field (Fig. 5b). In this case, specific heat in the quantum critical region varies according to the expression $C/T = \gamma_0 + a''T^{0.5}$ and the electrical resistance is described by the formula $\rho = \rho_0 + A''T^{1.5}$. The authors of paper [29] believe that the magnetic field has a substantial effect on the character of fluctuations in the compound $CeCu_{5.8}Au_{0.2}$ at a QCP. We note, however, that the shape of the phase transition line $T_N(P)$ contradicts the Nernst heat theorem, and it therefore can not be excluded that a more accurate determination of the QCP coordinate may affect the conclusions drawn in [29].

(ii) As was shown in paper [30], the heavy-fermion antiferromagnet $YbRh_2Si_2$ can be transformed into a paramagnetic state by a comparatively small magnetic field $H_c \approx 0.6$ T (Fig. 6). Electrical resistance in the antiferromagnetic state is best described by the expression $\Delta\rho = AT^2$ with a very large coefficient $A = 22$ μΩ cm K^{-2} in the region $20 \leqslant T \leqslant 60$ mK and $H = 0$. In a magnetic field $H = H_c$ the temperature dependence of the resistance follows the linear law up to the lowest attainable temperatures (20 mK). For $H > H_c$ and $T < T^*$ (see Fig. 6), electrical resistance is again

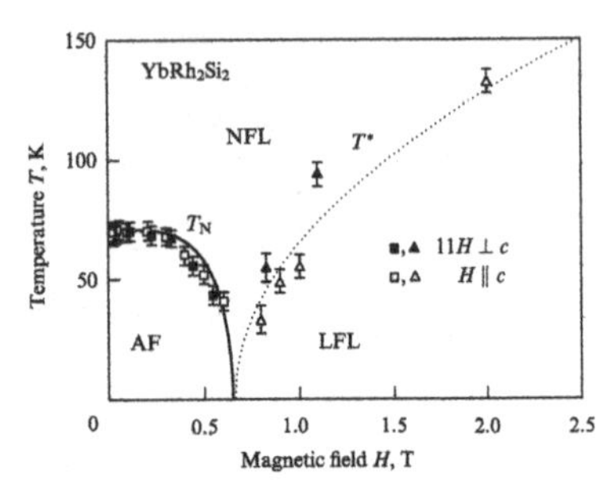

Figure 6. Phase diagram of $YbRh_2Si_2$ in a magnetic field [30]. T_N is the Neel temperature and T^* is the boundary of Fermi-liquid behavior. The values of the corresponding quantities in the perpendicular field are multiplied by 11. AF is an antiferromagnet, NFL is a non-Fermi liquid, and LFL is a Landau Fermi liquid.

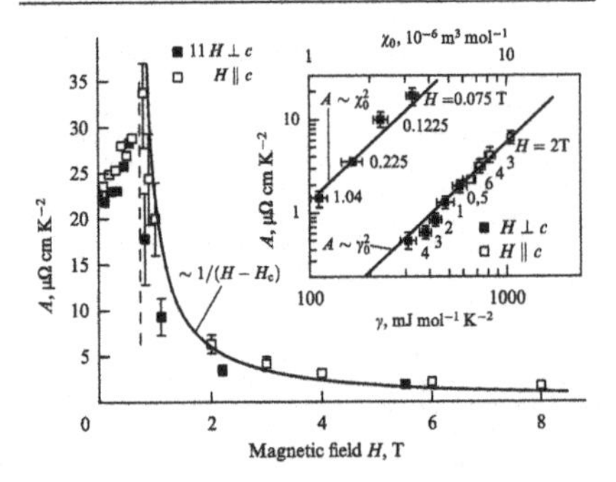

Figure 7. Dependence of the coefficient $A = \Delta\rho/T^2$ on the magnetic field for $YbRh_2Si_2$ [30]. The data for $H \perp c$ are multiplied by 11. The dashed line marks the H_c values, the solid curve corresponds to the dependence $\sim 1/(H - H_c)$. The inset gives the dependencies $A(\gamma_0)$ and $A(\chi_0)$ on a double logarithmic scale.

described by the Fermi-liquid expression $\Delta\rho = AT^2$. Being proportional to the electron-electron scattering cross section, the coefficient A diverges as $A(H) \propto 1/(H - H_c)$ when $H \to H_c$ (Fig. 7). An analysis of the behavior of electrical resistance and specific heat in longitudinal and transverse magnetic fields suggests the quasiparticle effective mass divergence $1/(H - H_c)^{1/2}$ when $H \to H_c$. This conclusion was confirmed in a study of the doped compound $YbRh_2(Si_{0.95}Ge_{0.05})_2$ in weak magnetic fields [31]. In this case, a small germanium impurity expands the lattice to shift the magnetic coordinate of the QCP nearer to the value $H = 0$, which makes it possible to check whether any new property appears in a QCP obtained in a strong magnetic field. In the particular case of $YbRh_2(Si,Ge)_2$ the properties of QCPs with coordinates $H_c \approx 0$ and $H_c \neq 0$ turn out to be identical. In both compounds in the quantum critical region for $H = H_c$ the magnetic susceptibility behaves like $\chi^{-1} \propto T^\alpha$, where $\alpha \approx 0.75$ (0.3–1.5 K), and the coefficient of electronic specific heat C_{el}/T diverges logarithmically at

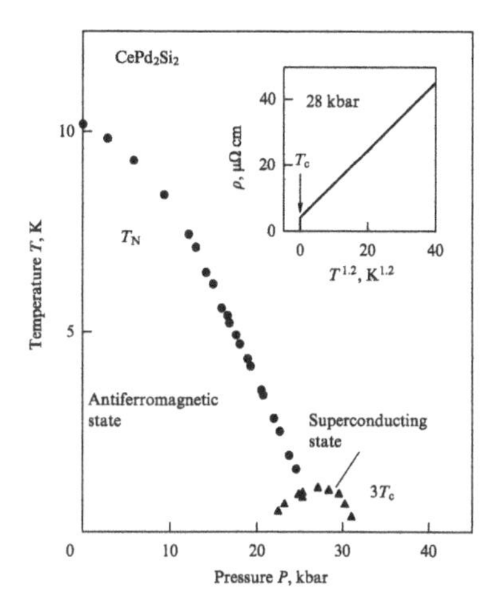

Figure 8. Phase diagram of $CePd_2Si_2$ [32]. Superconductivity occurs in a narrow pressure range in the region where the Neel temperature T_N tends to zero. For clarity, the superconducting transition temperatures T_c are multiplied by 3. The inset illustrates the non-Fermi-liquid behavior of $CePd_2Si_2$ electrical resistance for $P = 28$ kbar.

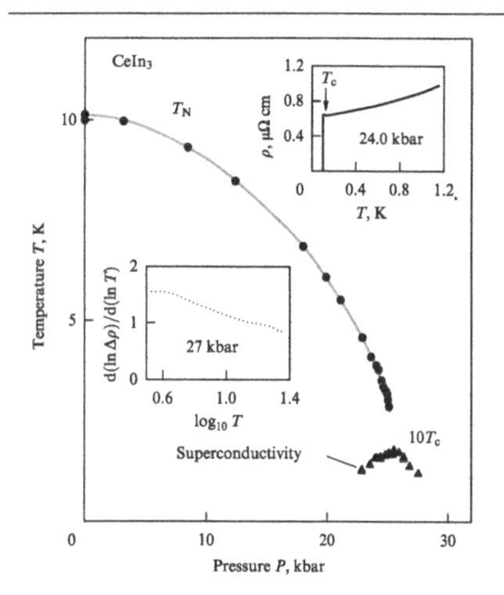

Figure 9. Phase diagram of $CeIn_3$ [32]. The upper inset shows that the superconducting transition is completed even at a pressure below critical. The lower inset demonstrates that the electrical resistance in the normal state varies as $T^{1.6\pm0.2}$ at a pressure above critical. The superconducting transition temperatures are multiplied by 10.

temperatures between 0.3 and 10 K. The electrical resistance exhibits a linear temperature dependence in the range of 0.02 to 0.5 K.

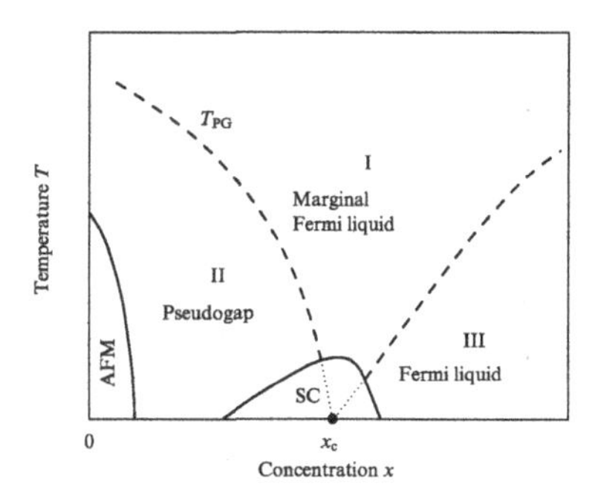

Figure 10. Generalized phase diagram of superconducting cuprates with hole doping [9]. T_{PG} is the boundary of the pseudogap state. AFM is the antiferromagnetic phase, SC is the superconducting phase, x_0 is a virtual quantum critical point. The properties of cuprates in region I resemble the properties of heavy-fermion metals in the quantum-critical region.

(iii) $CePd_2Si_2$, and $CeIn_3$ [32] are heavy-fermion compounds with an intriguing phase diagram (Figs 8, 9). One can see from the figures that the phase diagrams of these compounds are similar. The antiferromagnetic phase transition temperature T_N drops in both cases as pressure increases. In both cases, the phase transition curve should have crossed the pressure axis with the formation of a QCP crowning the antiferromagnetic-paramagnetic transition. However, at low temperatures both curves actually end at the top of the 'superconducting dome', which evidently points to the genetic relation between quantum critical phenomena and superconductivity in these magnetic systems. In this connection it is of interest to compare the phase diagrams of Figs 8 and 9 with the generalized phase diagram of superconductive cuprates (Fig. 10) [9]. Isomorphism of the phase diagrams becomes obvious if we assume the pseudogap line T_{PG} to be in a sense equivalent to the magnetic phase transition curve.

It is noteworthy that the temperature dependence of the electrical resistance of both substances in the quantum critical region does not correspond to Fermi-liquid behavior.

4.2 Itinerant ferromagnets UGe_2 and $ZrZn_2$

The properties of the ferromagnetic compound UGe_2 [33–35] represent another interesting example of the relation among quantum critical phenomena, magnetism, and superconductivity. As Fig. 11 shows, a small superconducting domain in this case lies entirely in the region of ferromagnetic phase stability near the phase boundary. It is of importance that, as established with the help of elastic scattering of neutrons, ferromagnetic ordering is preserved in the transition to a superconducting state. At low temperatures, a ferromagnetic phase transition becomes a first-order phase transition, and therefore the role of quantum fluctuations in the possible formation of a triplet superconducting state in UGe_2 seems to be vague. Useful information in this respect can be gained from the phase diagram of another ferromagnetic superconductor, $ZrZn_2$ [36] (Fig. 12). The relation between ferromagnetism and

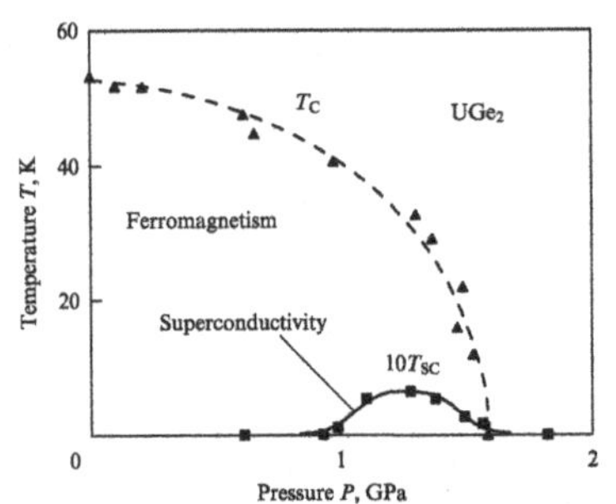

Figure 11. Phase diagram of UGe_2 [33]. T_C is the Curie temperature, T_{SC} is the superconducting transition temperature (T_{SC} values are multiplied by 10).

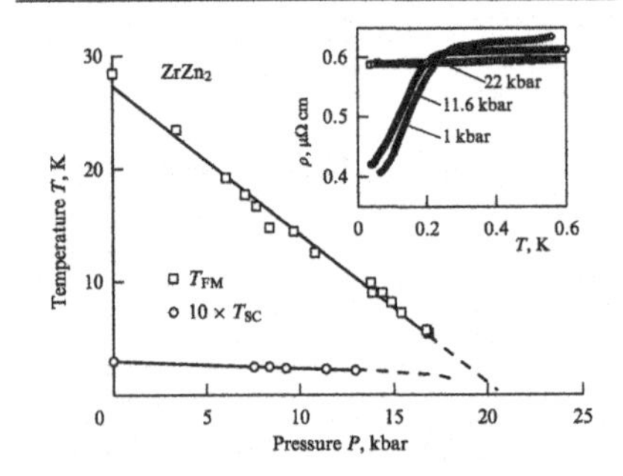

Figure 12. Pressure dependence of ferromagnetic (T_{FM}) and superconducting (T_{SC}) transition temperatures in $ZrZn_2$ [36] (T_{SC} values are multiplied by 10).

superconductivity seems obvious in this case [11], but the quantum critical fluctuations have apparently nothing to do with this.

With this it seems expedient to conclude the present review, which may to some extent be called an introduction to the subject. Its goal is to draw the attention of Russian researchers to an important and rapidly developing trend in the physics of strongly correlated systems. Although not exhaustive, the list of references includes the most important original studies and reviews. The present survey is based on a talk given at the seminar 'Strongly correlated electron systems and quantum critical phenomena' held on April 11, 2003 in Troitsk.

References

1. Sachdev S *Quantum Phase Transitions* (Cambridge: Cambridge Univ. Press, 1999)
2. Continentino M A *Quantum Scaling in Many-Body Systems* (Singapore: World Scientific, 2001)
3. Continentino M A *Phys. Rep.* **239** 179 (1994)
4. Sondhi S L et al. *Rev. Mod. Phys.* **69** 315 (1997)
5. Vojta T *Ann. Phys.* (Leipzig) **9** 403 (2000)
6. Coleman P et al. *J. Phys.: Condens. Matter* **13** R723 (2001)
7. Stewart G R *Rev. Mod. Phys.* **73** 797 (2001)
8. Lavagna M *Philos. Mag. B* **81** 1469 (2001)
9. Varma C M, Nussinov Z, van Saarloos W *Phys. Rep.* **361** 267 (2002)
10. Hertz J A *Phys. Rev. B* **14** 1165 (1976)
11. Millis A J *Phys. Rev. B* **48** 7183 (1993)
12. Pfeuty P, Elliott R J *J. Phys. C: Solid State Phys.* **4** 2370 (1971)
13. Young A P *J. Phys. C: Solid State Phys.* **8** L309 (1975)
14. Landau L D, Lifshitz E M *Statisticheskaya Fizika* (Statistical Physics) Pt. 1 (Moscow: Fizmatlit, 1995) [Translated into English (Oxford: Pergamon Press, 1980)]
15. Patashinskiĭ A Z, Pokrovskiĭ V L *Fluktuatsionnaya Teoriya Fazovykh Perekhodov* (Fluctuation Theory of Phase Transitions) 2nd ed. (Moscow: Fizmatlit, 1982) [Translated into English of 1st Russian ed. (Oxford: Pergamon Press, 1979)]
16. Stanley H E *Introduction to Phase Transitions and Critical Phenomena* (Oxford: Clarendon Press, 1971) [Translated into Russian (Moscow: Mir, 1973)]
17. Levanyuk A P *Zh. Eksp. Teor. Fiz.* **36** 810 (1959) [*Sov. Phys. JETP* **9** 571 (1959)]
18. Ginzburg V L *Fiz. Tverd. Tela* **2** 2034 (1960) [*Sov. Phys. Solid State* **2** 1824 (1961)]
19. Goldenfeld N *Lectures on Phase Transitions and the Renormalization Group* (Reading, Mass.: Addison-Wesley, Adv. Book Program, 1992)
20. Cardy J *Scaling and Renormalization in Statistical Physics* 2nd ed. (Cambridge: Cambridge Univ. Press, 2002)
21. Erkelens W A et al. *Europhys. Lett.* **1** 37 (1986)
22. Andraka B, Tsvelik A M *Phys. Rev. Lett.* **67** 2886 (1991)
23. Aronson M C et al. *Phys. Rev. Lett.* **75** 725 (1995)
24. Tsvelik A M, Reizer M *Phys. Rev. B* **48** 9887 (1993)
25. Zhu L et al. *Phys. Rev. Lett.* **91** 066404 (2003)
26. Hewson A C *The Kondo Problem to Heavy Fermions* (Cambridge: Cambridge Univ. Press, 1993)
27. Pietrus T et al. *Physica B* **206–207** 317 (1995)
28. Schröder A et al. *Nature* **407** 351 (2000)
29. Stockert O et al. *Physica B* **312–313** 458 (2002)
30. Gegenwart P et al. *Phys. Rev. Lett.* **89** 056402 (2002)
31. Custers J et al. *Nature* **424** 524 (2003)
32. Mathur N D et al. *Nature* **394** 39 (1998)
33. Saxena S S et al. *Nature* **406** 587 (2000)
34. Settai R et al. *J. Phys.: Condens. Matter* **14** L29 (2002)
35. Sandeman K G, Lonzarich G G, Schofield A J *Phys. Rev. Lett.* **90** 167005 (2003)
36. Pfleiderer C et al. *Nature* **412** 58 (2001)
37. Stishov S M et al. (to be published)

[11] The recent study of the phase diagram of $ZrZn_2$ [37] is apparently indicative of the absence of a direct relation between superconductivity and magnetism in this compound because a superconducting phase transition is also observed in the paramagnetic phase stability region.

A.8. Phase transition in helical magnet MnSi

A.8.1. Itinerant helimagnet MnSi

Physics – Uspekhi **54** (11) 1117–1130 (2011)

REVIEWS OF TOPICAL PROBLEMS

PACS numbers: **62.50.–p**, 75.30.Kz, 75.40.Cx, 77.80.B–

Itinerant helimagnet MnSi

S M Stishov, A E Petrova

DOI: 10.3367/UFNe.0181.201111b.1157

Contents

Abstract. Manganese silicide (MnSi), a model helimagnetic compound, crystallizes in a B20 structure, whose non-centrosymmetric space group $P2_13$ allows a helical (chiral) magnetic structure. The magnetic phase transition temperature of MnSi (29 K at atmospheric pressure) decreases with pressure and approaches zero at about 1.4 GPa. This fact, pointing to possible quantum critical phenomena, has prompted high-pressure studies of MnSi that have revealed a number of fascinating phenomena, in particular, the non-Fermi-liquid behavior of electrical resistivity and the unusual spin state (partial ordering) in the paramagnetic phase. We discuss experimental results characterizing the physical properties and the phase diagram of MnSi and at the same time forming the basis of current ideas in the physics of strongly correlated chiral electron systems.

1. Introduction

The intermetallic compound MnSi (manganese silicide) belongs to the class of weak itinerant-electron (band) magnets with a low Curie temperature and small magnetic moment (a few fractions of the Bohr magneton). This class also includes compounds such as FeGe, $Fe_{1-x}Co_xSi$, $ZrZn_2$, Ni_3Al, Ni_3Y, Ni_3Ga, Sc_3Al, and CoS_2. The magnetic susceptibility of these compounds in the paramagnetic phase approximately obeys the Curie–Weiss law, with the effective magnetic moment in the paramagnetic phase significantly exceeding the magnetic moment derived from the saturation magnetization.

In contrast to other substances mentioned above, MnSi, FeGe, and $Fe_{1-x}Co_xSi$ compounds are helimagnets (have a helical magnetic structure), which to a significant extent determines the variety of observed physical phenomena and separates these materials from the class of weak band magnets. But because MnSi has been studied the most of all these compounds (which is largely due to the availability of high-quality single crystals), it is the description of its properties that became the subject of numerous theories and concepts stimulating the development of new experiments. In fact, MnSi is a model substance that allows studying the influence of the weak chiral Dzyaloshinskii–Moriya interaction on the physics of strongly correlated systems.

MnSi, as well as a number of other silicides and germanides of transition metals (FeSi, CoSi, FeGe, etc.), crystallizes into the structure type B20 (space group $P2_13$) [1]. This space group has no inversion center, which allows a helical (chiral) magnetic structure to exist in these compounds (which indeed have been revealed in MnSi, $Fe_{1-x}Co_xSi$, and FeGe).

We note that the specific features of the magnetic structure of these compounds significantly favored the development of adequate experimental and theoretical methods of neutron diffraction studies.

The magnetic phase transition temperature in MnSi (about 29 K at atmospheric pressure) decreases with increasing pressure and tends to zero at the pressure ≈ 1.4 GPa [2]. This circumstance, indicating the possibility of observing quantum critical phenomena, determined the subsequent interest in the investigations of MnSi at high pressures, which have revealed a number of intriguing features in the behavior of MnSi. A deviation from the Fermi-liquid behavior was discovered in the paramagnetic phase [3, 4]. An unusual spin state (partial order) was found in the

S M Stishov, A E Petrova Vereshchagin Institute for High Pressure Physics, Russian Academy of Sciences, Kaluzhskoe shosse 14, 142190 Troitsk, Moscow region, Russian Federation
E-mail: *sergei@hppi.troitsk.ru*

Received 28 February 2011, revised 6 April 2011
Uspekhi Fizicheskikh Nauk **181** (11) 1157–1170 (2011)
DOI: 10.3367/UFNr.0181.201111b.1157
Translated by S N Gorin; edited by A M Semikhatov

paramagnetic phase [5]. To explain these phenomena, new concepts were suggested, such as skyrmions [6] and a "helical spin crystal" [7] (also see [8]). Noteworthy is the interpretation of the results of small-angle neutron scattering (SANS) in the so-called A phase of MnSi in terms of a skyrmion crystal [9]. Here, we should keep in mind that the skyrmions in magnetic substances are vortex spin structures, which in some cases can 'crystallize' to form vortex lattices. The term 'skyrmion' is related to the name of the English physicist Tony Skyrme, who suggested a topological model of baryons (see [6] and the references in Appendix I).

The physical properties of MnSi have been under intense investigations for several decades at ambient and high pressures, which has led to a number of new results that are, in our opinion, of great importance for all of physics and, consequently, are of interest for a wide circle of readers of this journal.

It was initially planned that this review would be written together with S V Maleev and S V Grigor'ev (Konstantinov Institute of Nuclear Physics, St. Petersburg), who are the authors of a series of extensive studies on neutron scattering in MnSi. However, this did not happen and, in publishing this review, we expect that their review will also appear soon. For this reason, we virtually ignore some issues that require a qualified analysis of the results of neutron investigations. Nor do we consider some studies devoted to investigations using nuclear magnetic resonance and electron paramagnetic resonance methods or studies that do not lead to fundamentally new concepts.

Below, as the characteristics of the magnetic field, we use two quantities, the magnetic field strength H expressed in oersteds, and the magnetic induction B expressed in teslas ($1\ \mathrm{T} = 10^4$ Oe).

2. General characteristic of MnSi

MnSi is a weak itinerant-electron (band) ferromagnetic 3d metal that crystallizes into a B20-type noncentrosymmetric cubic structure characterized the space group $P2_13$ containing no inversion center. The primitive unit cell with the lattice parameter $a \approx 4.56$ Å contains four formula units.

The B20 structure type includes right-handed and left-handed enantiomorphic forms (Fig. 1). The coordinates of Mn and Si atoms of the right-hand form are:

$$(x,x,x,),\left(\frac{1}{2}+x,\frac{1}{2}-x,-x\right),\left(-x,\frac{1}{2}+x,\frac{1}{2}-x\right),$$
$$\left(\frac{1}{2}-x,-x,\frac{1}{2}+x\right),$$

where x takes the values $x_{\mathrm{Mn}} = 0.137$ and $x_{\mathrm{Si}} = 0.845$. The left-hand form corresponds to the replacement of x with $1 - x$ (see Fig. 1) [10, 11].

MnSi is a congruently melting compound and its single crystals can be obtained both by direct crystallization from the melt using Bridgman and Czochralski techniques, floating zone method, and so on, and by crystallization from the solution in the melt of some metals, e.g., tin. A magnetically ordered state in MnSi was discovered in [12], where it was also shown that the magnetic ordering in MnSi occurs at the temperature 30 K. In [13], a helical magnetic order was found in MnSi.

The magnetic structure of MnSi in a zero magnetic field can be described as a system of ferromagnetically ordered planes located parallel to the (111) crystallographic plane.

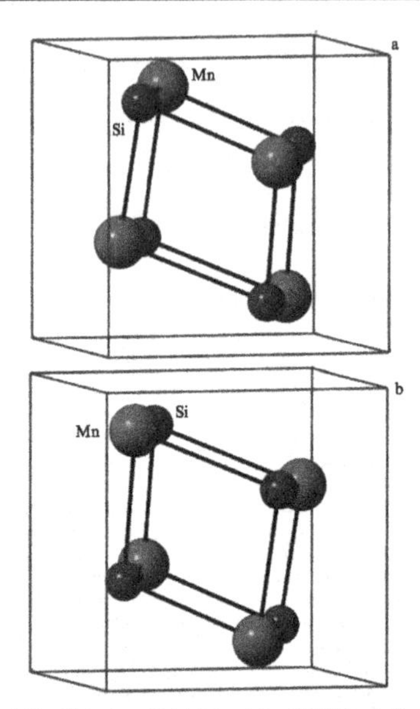

Figure 1. Crystal structure of (a) right-handed and (b) left-handed forms of MnSi.

The magnetic moment of each layer is turned by a small angle relative to the magnetic moment of adjacent layers due to the Dzyaloshinskii–Moriya interaction $D[\mathbf{S}_1 \times \mathbf{S}_2]$, which is nonzero in noncentrosymmetrical structures [14–16]. As a result, in the magnetically ordered phase (at temperatures less than ≈ 29 K), the spins form a left-handed incommensurate spiral with the wave vector ≈ 0.036 Å^{-1} (corresponding to the period 180 Å) in the [111] direction. We note that the relation between the structural and magnetic chirality has not so far been completely clarified [10, 11, 17].

The magnetic properties of MnSi are controlled by a hierarchy of three characteristic energies, including the isotropic exchange, the weak Dzyaloshinskii–Moriya interaction, and the even weaker anisotropic exchange, which successively determine the ferromagnetic ordering in the layers, the formation of the spiral (chiral) structure, and the direction of the wave vector of the spiral.

It is interesting to compare the properties of MnSi with the properties of the silicides of the nearest neighbors of Mn in the periodic system, i.e., Fe and Co. The FeSi and CoSi compounds, just as MnSi, crystallize into the B20 structure type. The lattice parameters in the MnSi–FeSi–CoSi row change quite weakly as the atomic radii of the metallic elements change (Table 1). The elastic properties of these compounds at low temperatures change monotonically in accordance with the variation of the lattice parameters (Table 2).

The lattice properties of MnSi, FeSi, and CoSi appear to depend on the number of electrons on the 3d shell of metallic elements quite weakly. However, the same circumstance has a dramatic effect on the electrical and magnetic properties of

Table 1. Lattice parameters a and chemical composition of helimagnets [18].

Compound*	a, Å	Atomic percent	
		Metal	Si
MnSi	4.5598(2)	50.63	49.37
FeSi	4.4827(1)	50.16	49.84
CoSi	4.444(1)	50.07	49.93
	4.4438(6)		

* The MnSi, FeSi, and CoSi compounds belong to the so-called berthollides, i.e., compounds of variable composition. The exact values of the lattice parameters and the chemical composition of these compounds depend on their history.

Table 2. Elastic constants C_{ij} of MnSi, FeSi, and CoSi [18] (Θ_D is the Debye temperature).

	MnSi, $a = 4.5598$ Å, $\Theta_D = 660$ K		
C_{ij}	$T = 6.5$ K	$T = 78$ K	$T = 115$ K
C_{11}	3.2057	3.2045	3.2047
C_{44}	1.2615	1.2582	1.2540
C_{12}	0.8523	0.8574	0.8477
	FeSi, $a = 4.483$ Å, $\Theta_D = 680$ K		
C_{ij}	$T = 6.5$ K	$T = 77.8$ K	$T = 292.8$ K
C_{11}	3.4626	3.4454	3.1670
C_{44}	1.3916	1.3858	1.2521
C_{12}	1.0576	1.0608	1.1298
	CoSi, $a = 4.444$ Å, $\Theta_D = 625$ K		
C_{ij}	$T = 6.5$ K	$T = 77.8$ K	$T = 292.8$ K
C_{11}	3.5432	3.5404	3.4529
C_{44}	1.1847	1.1857	1.1593
C_{12}	1.3323	1.3331	1.3128

these compounds. Indeed, MnSi is a helical metallic magnet, FeSi is the so-called strongly correlated semiconductor with an unusual behavior of the magnetic susceptibility, and CoSi is a diamagnetic semimetal. The calculations of the energy band structures and density of states agree well with experimental data [19–21].

The magnetic moment of MnSi at $T = 1.4$ K calculated from the saturation magnetization in a magnetic field $H_c > 6.2$ kOe is $0.4\,\mu_B$ per Mn atom, whereas the approximation of the magnetic susceptibility in the paramagnetic phase by the Curie–Weiss relation yields the effective magnetic moment $2.2\,\mu_B$ per Mn atom [22]. Differences of this kind are considered to be a characteristic feature of band magnets.

Measurements of the thermal expansion and heat capacity of MnSi have demonstrated the existence of clearly pronounced anomalies in the vicinity of the magnetic phase transition, but did not lead to a definite conclusion on the nature of the phase transition [23]. The extrapolation of the C_p/T ratio to $T = 0$ gives an anomalously high value of the electronic heat capacity coefficient $\gamma \approx 36$ mJ mol^{-1} K^{-2}. The temperature dependence of the electrical resistivity at low temperatures is described well by a power law with an exponent close to 2, which indicates the Fermi-liquid behavior of the electron liquid in MnSi at ambient pressure. The electrical resistivity demonstrates a distinct anomaly in the region of the phase transition and tends to saturation at temperatures exceeding T_c [24, 25].

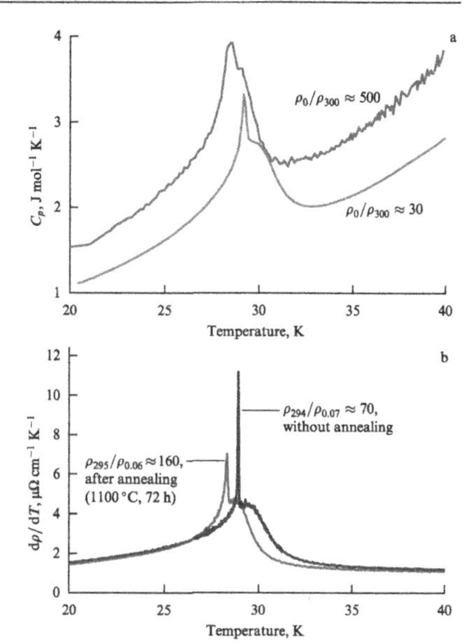

Figure 2. Effect of high-temperature annealing on (a) the heat capacity of MnSi (according to [26]) and (b) the temperature coefficient of resistivity of MnSi (V N Krasnorusskii and V N Narozhnyi, private communication).

In Sections 3 and 4, we describe new results of investigations of the physical properties of MnSi, which in some cases are based exclusively on our work performed on high-quality single crystals of MnSi.

Here, it is worth touching on the problem of the criterion of quality of crystals. In low-temperature investigations of metals, the magnitude of the ratio of the electrical resistivity at room temperature to the residual resistivity (at zero temperature) ρ_{295}/ρ_0 is usually chosen as such a criterion. We used just this criterion to estimate the quality of MnSi crystals. To increase ρ_{295}/ρ_0, crystals are frequently subjected to annealing for many hours at temperatures above 1000 °C. However, it turns out that this procedure adversely affects the characteristics of the phase transition (Fig. 2). Obviously, the ρ_{295}/ρ_0 criterion does not 'work' in the case of MnSi. On the other hand, the 'sharpness' of the phase transition can apparently serve as a qualitative indicator of the crystal perfection.

3. Physical properties of MnSi at ambient pressure

3.1 Magnetism

The most important characteristic of a magnetic substance is its response to an external magnetic field. The results of practically the first measurements have demonstrated an unusual behavior of the magnetic properties of MnSi (Figs 3, 4). The saturation magnetic moment of MnSi

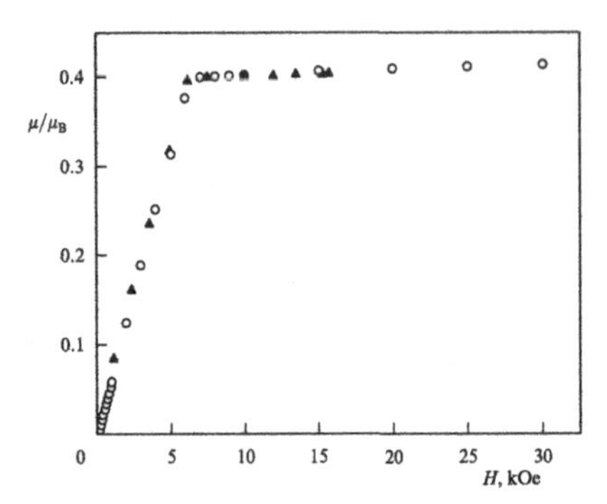

Figure 3. Saturation magnetic moment for MnSi: (▲) [22]; (○) [27].

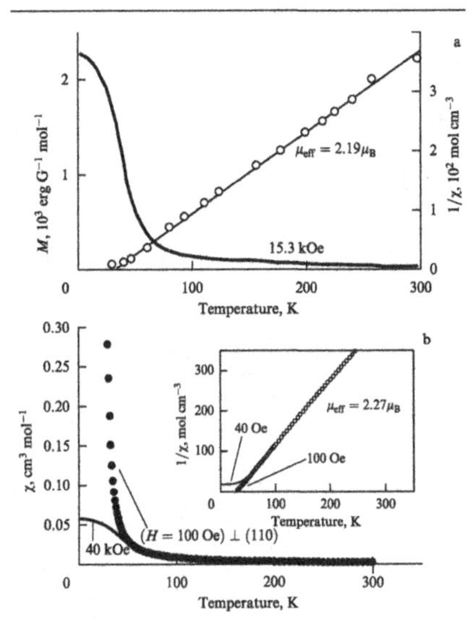

Figure 4. Magnetization and magnetic susceptibility of MnSi: (a) according to [22]; (b) according to [27] (the results corresponding to $H = 40$ kOe were obtained by J Thomson).

calculated from measurements in [22] was equal to 0.4 μ_B per Mn atom at $T = 1.4$ K. At the same time, the effective magnetic moment following from the Curie–Weiss behavior of the magnetic susceptibility of MnSi at $T > T_c$ is 2.19 μ_B per Mn atom. Later, this property of the magnetic characteristics of MnSi was assumed to be an intrinsic property of weak band magnets.

Figure 4b illustrates the results of recent measurements of the magnetic properties of MnSi obtained using a high-quality single crystal [27] in comparison with earlier data [22]. We note that the magnetization curves given in

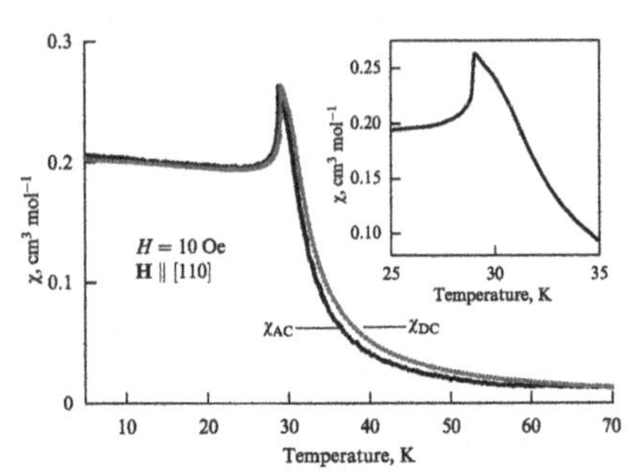

Figure 5. Magnetic susceptibility of MnSi in the vicinity of the phase transition [27] (χ_{AC} and χ_{DC} are the magnetic susceptibilities measured in an AC field and in a DC field).

Fig. 4 were measured in the magnetic fields 15.3 and 40 kOe in the field of the induced ferromagnetic phase, and do not therefore contain any features indicating the existence of a phase transition (see the phase diagram in Fig. 6). Magnetic measurements in fields less than 6 kOe demonstrate a clearly pronounced anomaly in the magnetic susceptibility at $T \approx 29$ K, corresponding to the phase transition temperature (Fig. 5). In some studies, this feature of the magnetic susceptibility is called the λ anomaly, which, as a matter of fact, does not correspond to the reality. The shape of the anomaly shown in the inset in Fig. 5 can be regarded as a result of a jump-like increase in the magnetic susceptibility at the phase transition point. In the absence of such a jump, the magnetic susceptibility curve would look like the standard $\chi(T)$ dependence characteristic of a typical antiferromagnetic phase transition.

Figure 6 displays the magnetic phase diagram of MnSi according to the data in [28–30]. It follows from this diagram that the application of a magnetic field first transforms the helical spin structure into a conical structure and then orients all spins in one direction, forming a field-induced ferromagnetic phase.

It is noteworthy that in magnetic fields less than 4 kOe, the phase transition temperature only weakly depends on the field (see Fig. 6), which appears to indicate a finite magnitude of the order parameter. This fact speaks in favor of the interpretation of the magnetic phase transition in MnSi as a first-order phase transition. Some difference in the behavior of the phase boundary as a function of the magnetic field in Figs 6a and 6b does not change the essence of this conclusions.

Further investigations of the magnetic phase diagram of MnSi led to the discovery of the so-called A phase, which forms a 'pocket' in the $H-T$ region of stability of the helical phase (Fig. 6a). The physical nature of the A phase is the subject of an ongoing discussion, although solid evidence in favor of its skyrmion nature has appeared recently [9]. We return to the discussion of this question below (see Appendix I).

3.2 Heat capacity

The heat capacity of MnSi was measured in crystals of various perfection in a number of studies using different techniques [23, 26, 31]. The first measurements of the heat capacity of

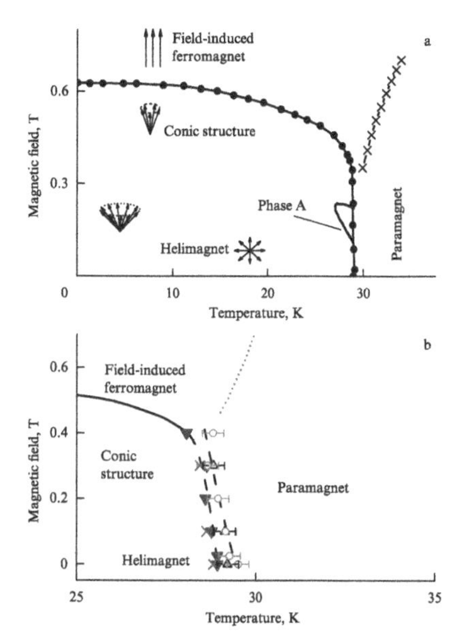

Figure 6. Magnetic phase diagram of MnSi: (a) according to ultrasonic investigations (● [28]) and the A phase according to [29]; (b) according to the measurements of thermal expansion (▼) and heat capacity (○) [30]. It is seen that in (b), in contrast to the data in [28] given in (a), the boundary between the helimagnetic and paramagnetic phases as a function of the magnetic field is tilted.

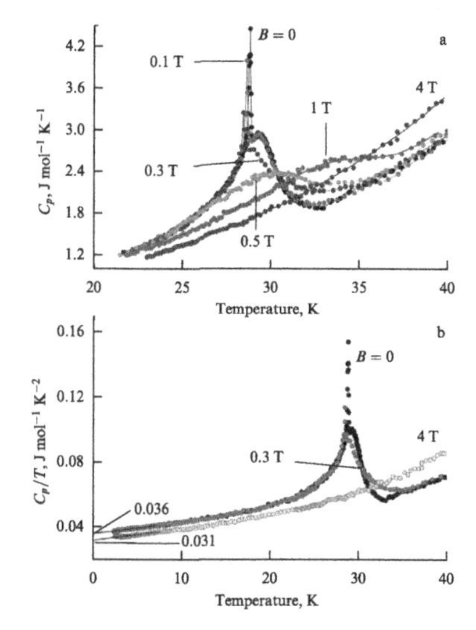

Figure 7. Heat capacity of MnSi in magnetic fields: (a) heat capacity in the phase transition region; (b) C_p/T ratio as a function of temperature.

MnSi were carried out (on samples of apparently moderate quality) in [23]. Nevertheless, by extrapolating the C_p/T ratio T to zero, Fawcett obtained a correct (according to the modern data) value of the electronic heat capacity coefficient $\gamma = 0.038$ J mol^{-1} K^{-2}. But an important feature (the sharp peak) in the heat capacity curve of MnSi in the phase transition region was not revealed. As a result, it was assumed in the physics community that the phase transition in MnSi should be regarded as a second-order transition in spite of the conclusions of the experimental and theoretical work [28, 29, 32].

New measurements of heat capacity were performed on high-quality crystals in magnetic fields up to 4 T (Fig. 7) [27, 30]. It can be seen that the temperature dependence of the heat capacity $C_p(T)$ at $B = 0$ is characterized by a sharp peak at $T \approx 28.8$ K, corresponding to the phase transition with a clearly pronounced 'shoulder' on its high-temperature side. An increase in the magnetic field to 0.3 T only insignificantly decreases the phase transition temperature and leaves the heat capacity of the magnetically ordered phase virtually unaffected, which apparently indicates a certain hardness of the magnetic structure and, consequently, a finite magnitude of the order parameter at the phase transition point in MnSi. The magnetic field 0.5 T transforms the system into a field-induced ferromagnetic state and eliminates the phase transition (see Fig. 6).

Figure 7b, which illustrates the behavior of the C_p/T ratio, allows determining the electron heat capacity coefficients as $\gamma = 36$ mJ mol^{-1} K^{-2} at $B = 0$ and $\gamma = 31$ mJ mol^{-1} K^{-2} at $B = 4$ T. The value of γ at $B = 0$ agrees well with the earlier estimate (in which the evident error in the order of magnitude was corrected) [23]. The high value of γ indicates a significant effective electron mass in MnSi. A rough estimate performed in [30] gives $m^*/m \approx 36$.

The obtained value of γ was used in [33] in the analysis of the behavior of the magnetic contribution to the heat capacity of MnSi, which revealed the existence of negative contributions to the heat capacity and entropy at $T > T_c$ (see Appendix II for the details).

To conclude this section, we note that both experimental and calculated data are currently available that allow reliably determining the phonon contribution to the heat capacity of MnSi (see [33] and Appendix II). Nevertheless, a complete analysis of the behavior of the heat capacity of MnSi, including the separation of the magnon contribution, is difficult in view of the existence of a large anomaly in the phase transition region. This anomaly cannot be described in terms of the standard models of phase transitions; its description requires a nontrivial approach. We also note that the data obtained in [34, 35] indicate the existence in MnSi of helical magnons with a specific dispersion law.

3.3 Thermal expansion

The heat expansion of MnSi was first measured in [23], where a wide negative anomaly of the thermal expansion coefficient was revealed in the vicinity of the phase transition. A complex structure of the anomaly was observed (apparently, also for the first time) in [36]. The anomaly (noted previously in the example of heat capacity) was characterized, apart from the

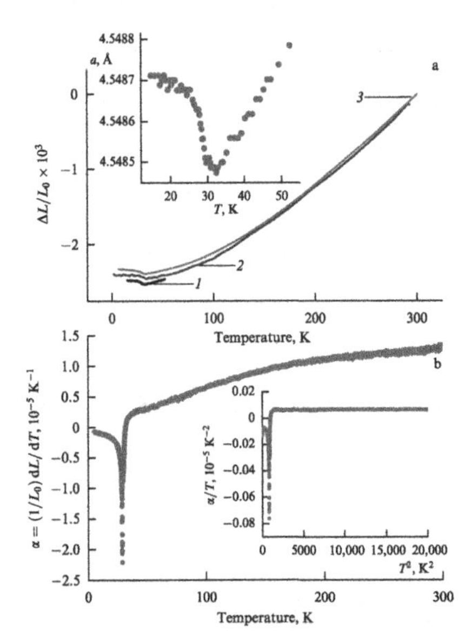

Figure 8. Thermal expansion of MnSi [27, 30]: (a) linear thermal expansion according to (*1*) XRD data, (*2*) neutron-diffraction data, and (*3*) dilatometric data; (b) thermal-expansion coefficient (the inset illustrates the linear behavior of the TEC).

main negative peak at the phase transition point, by a side minimum (shoulder) at a temperature approximately 2 K above T_c.

Detailed investigations of the thermal expansion of MnSi have been performed in [27, 30]; the results are given in Figs 8 and 9.

Figure 8a shows the behavior of the linear thermal expansion coefficient of MnSi according to the data obtained using capacitance dilatometry and X-ray and neutron diffraction. All the results are qualitatively consistent, although from the quantitative standpoint, preference should be given to the X-ray diffraction data. On the other hand, data on the behavior of the thermal expansion coefficient are mainly based on the results of dilatometric measurements, because neither X-ray nor neutron diffraction have sufficient resolution for the investigation of the fine details of thermal expansion of MnSi in the vicinity of the phase transition.

Figure 8b displays the temperature dependence of the linear thermal expansion coefficient $\alpha = (1/L_0)\,\mathrm{d}L/\mathrm{d}T$ of MnSi. It can be seen that the behavior of α in the temperature range 0–35 K is determined by magnetostrictive and fluctuation effects arising as a result of ordering of magnetic moments. The representation of the results as a dependence of α/T on T^2 does not reveal a noticeable lattice contribution to the thermal expansion of MnSi but shows a linear increase in the thermal expansion coefficient α in the temperature range 35–150 K (see the inset in Fig. 8b).

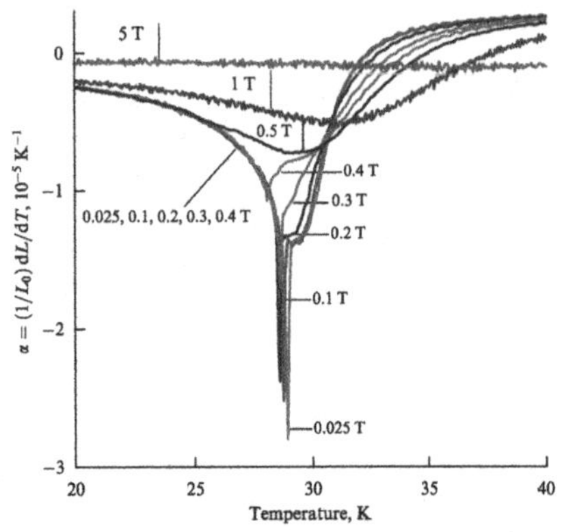

Figure 9. Thermal expansion coefficient of MnSi in the vicinity of the phase transition [27].

Figure 9 illustrates the effect of a magnetic field on the thermal expansion coefficient of MnSi. Again, as in the case of heat capacity, a moderate magnetic field (up to 0.4 T) leaves the thermal expansion of the helical phase virtually unaffected but substantially influences the behavior of the paramagnetic phase. We note that the side minimum (shoulder) vanishes faster than the main minimum corresponding to the phase transition. In any case, the behavior of the thermal expansion coefficient of MnSi confirms the earlier conclusion of a finite value of the order parameter at the phase transition point, which obviously contradicts the concept of its continuity. We also note that the fast degradation of the phase transition in magnetic fields between 0.4 and 0.5 T indicates the formation of a field-induced ferromagnetic spin structure in MnSi.

3.4 Elastic properties and ultrasound attenuation

In our previous work [37], we studied the elastic properties of MnSi and attenuation of ultrasound in it using the digital pulsed technique of measurements [38]. The results in [37] are presented graphically in Fig. 10 (selected numerical data are given in Table 2). As can be seen from Fig. 10a, the magnetic phase transition in MnSi is accompanied by a profound anomaly (softening) of elastic moduli and of their combinations that determine the propagation of longitudinal ultrasonic waves. At the same time, the magnetic phase transition manifests itself as a small change (jump) in the elastic moduli, which is localized on the low-temperature side of the anomalies. We note that both these features excellently correlate with the behavior of the heat capacity, the thermal expansion coefficient, and the thermal resistivity coefficient (see Section 3.5), except that the sharp maxima and minima of these thermodynamic and transport properties at the phase transition point are replaced by moderate jumps of the elastic moduli. It is precisely such a behavior of the elastic moduli that should be expected in the case of a weak first-order phase transition [37].

In contrast to the 'longitudinal' moduli, the shear moduli change insignificantly, mainly following the variation of the volume deformation arising upon magnetic ordering

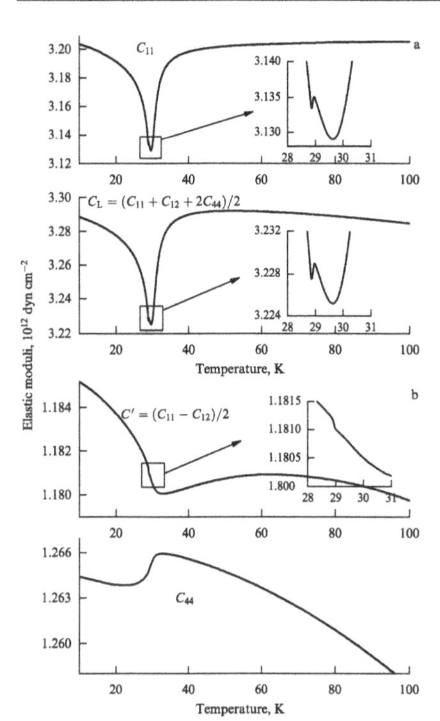

Figure 10. Variation of the elastic moduli during the phase transition in MnSi [37].

(Fig. 10b). This significant difference in the behavior of the longitudinal and transverse elastic characteristics appears to be related to the insignificant role of the interaction between the ionic and electronic subsystems during the magnetic phase transition in MnSi, simultaneously emphasizing the itinerant-electron nature of magnetism in this substance.

We now turn to Fig. 11, which illustrates the attenuation of ultrasound waves. It is interesting that the total structure of the attenuation curves (the main peak and the side maximum) represents an almost exact copy of the corresponding curves that characterize the behavior of the heat capacity, the thermal expansion coefficient (with the inverse sign), and the thermal resistivity coefficients (see Section 3.5). It is noted in [37] that the main peak in the attenuation curves appears to be due to the adiabaticity distortion caused by a finite change in the entropy under the first-order phase transition in MnSi.

The investigations in [37] also reveal a very important fact, which clearly demonstrates that the magnetic phase transition at $T = 28.8$ K in MnSi is only one of the features of a global transformation accompanied by the appearance of anomalies of physical quantities with extrema located at temperatures somewhat higher than the phase transition temperature. The nature of this transformation is still not completely clear.

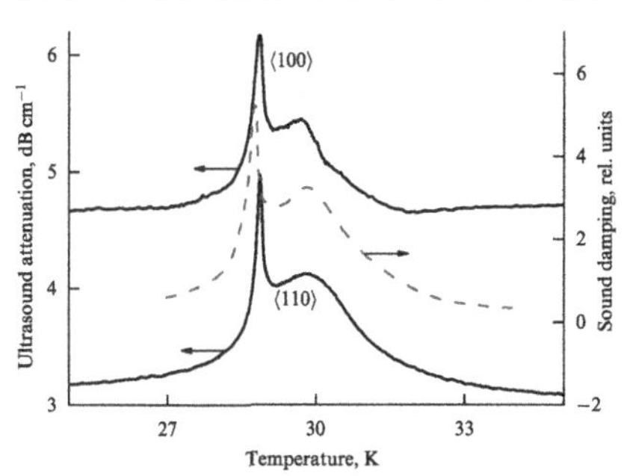

Figure 11. Ultrasound attenuation during the phase transition in MnSi: solid curves, data in [37]; dashed curve, data in [39].

3.5 Electrical resistivity

In this section, we only touch on the temperature behavior of the electrical resistivity of MnSi in the vicinity of the phase transition at ambient pressure according to [27]. As we can see from Fig. 12c, the resistivity of MnSi increases monotonically, demonstrating a tendency to saturation at temperatures above T_c. Just at the phase transition point, a small ($\approx 1.6\%$), slightly smeared jump is observed. The derivative $d\rho/dT$ exhibits a wide maximum with a sharp peak, which is associated with a smeared δ function. It is instructive to compare the overall shape of the $d\rho/dT$ function with the corresponding curves characterizing the behavior of heat capacity and thermal expansion (see Fig. 12).

3.6 Nature of the phase transition in MnSi

The order parameter in MnSi is the spin density (slowly changing in space) $\mathbf{S}(\mathbf{r}) = \mathbf{S}_1 \cos(\mathbf{qr}) - \mathbf{S}_2 \sin(\mathbf{qr})$, where $\mathbf{S}_1 \perp \mathbf{S}_2$ and $\mathbf{q}$ is the wave vector of the spiral. The expansion of the free energy of MnSi in powers of the spin density does not contain odd terms because of the symmetry under time reversal, which implies the possibility of the realization of the magnetic phase transition in MnSi as a second-order phase transition [32, 40]. Nevertheless, the renormalization-group analysis performed in [32] led the authors of [32] to the conclusion of the jump-like nature of the phase transition in MnSi induced by fluctuations. At that time, this conclusion was supported by the experimental data obtained in investigations of the magnetic phase diagram of MnSi [28, 29].

The modern data generalized in Fig. 12 demonstrate sharp peaks in the heat capacity curves, the thermal expansion coefficient, and the thermal resistivity coefficient, and slightly smeared jumps in the corresponding integrated curves, which allows interpreting the phase transition in MnSi as a weak first-order phase transition. Ultrasonic investigations, which revealed jumps in a number of elastic constants at the phase-transition point, agree well with this statement (Fig. 10a).

On the whole, however, the situation is by no means so simple: as we can see from Figs 11 and 12, in the curves that characterize the behavior of the heat capacity, the thermal expansion coefficient, the thermal resistivity coefficient, and ultrasound attenuation during the phase transition in MnSi, besides a sharp peak corresponding to the first-order phase

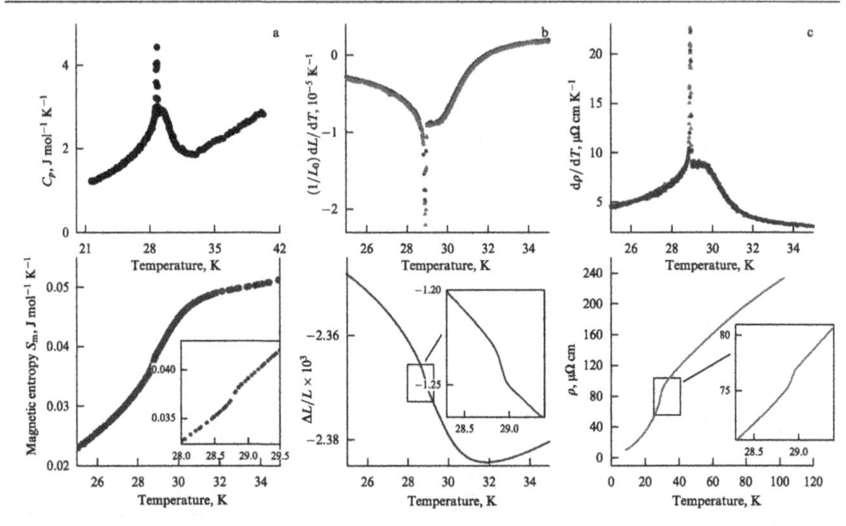

Figure 12. Behavior of thermodynamic quantities and the electrical resistivity during the phase transition in MnSi (according to [3, 27]). For clarity, in the inset given in (b), the background component has been subtracted from the total curve.

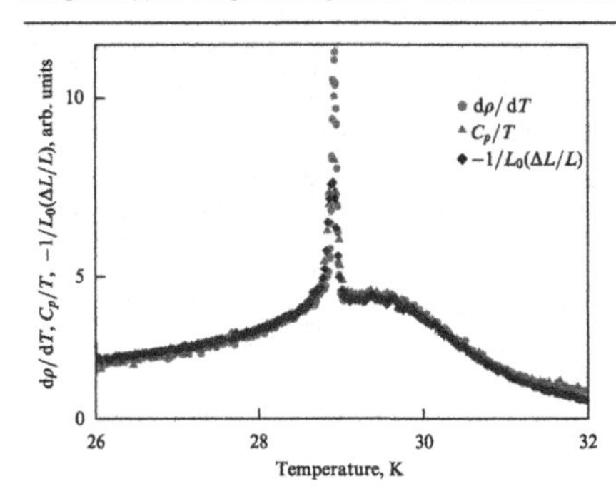

Figure 13. Reduced values of the heat capacity, the thermal expansion coefficient, and the thermal resistivity coefficient of MnSi in the vicinity of the phase transition [3].

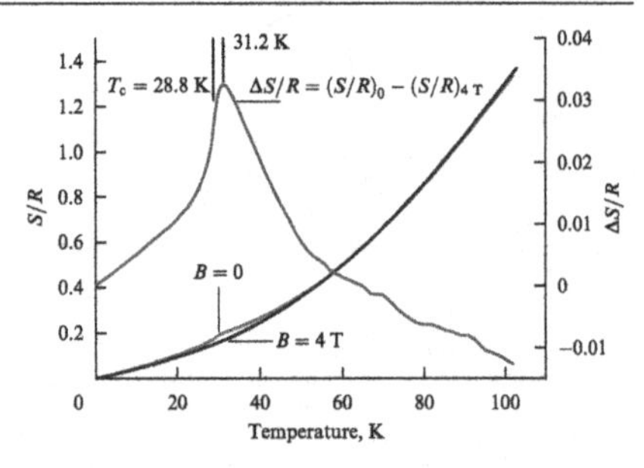

Figure 14. Entropy of MnSi calculated from the data on heat capacity measured at $B = 0$ and $B = 4$ T [30, 33]. The differential curve $\Delta S/R$ demonstrates that the drop in the entropy occurs at temperatures somewhat higher than the phase transition temperature.

transition, there also exist flat maxima or a minimum of the thermal expansion coefficient. It is surprising that the behavior of the above-mentioned quantities in the vicinity of the phase transition can be represented in the form of a single reduced curve (Fig. 13); still, this does not explain the nature of the flat extrema mentioned above.

Useful information can be obtained from Fig. 14, which demonstrates that a significant decrease in the entropy occurs beginning from a temperature of 31.2 K, close to the temperature of the corresponding extrema. In this connection, it is noteworthy to recall the ideas on the possible splitting of the phase transition in chiral magnetic systems into two transitions, magnetic and chiral, separated by a certain temperature interval [41, 42]. Papers [6, 43] should also be mentioned, in which the effects observed at $T > T_c$ are ascribed to the formation of skyrmions.

4. Phase diagram and investigations of MnSi at high pressures

4.1 Phase diagram

The first experiments on the effect of high pressure on the magnetic phase transition temperature T_c in MnSi were described in [3]. It was found that T_c decreases as the pressure increases, tending to zero at the pressure

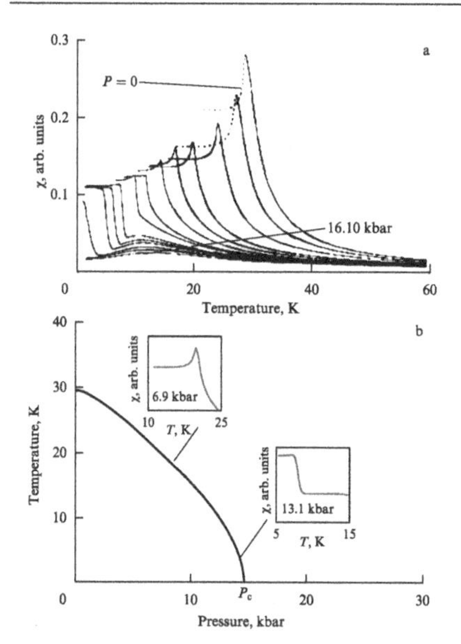

Figure 15. (a) Magnetic susceptibility and (b) phase diagram of MnSi according to [44, 45].

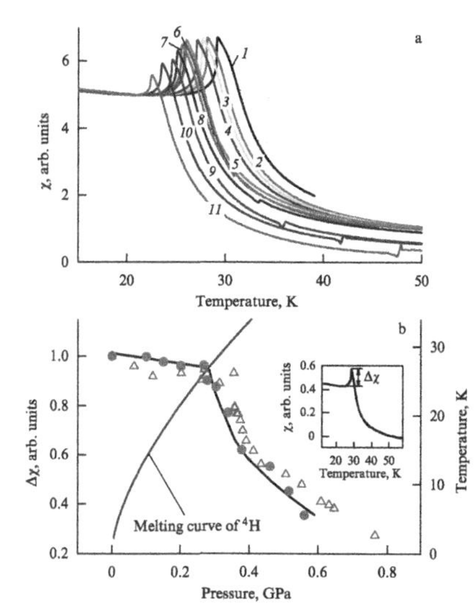

Figure 16. (a) Temperature dependence of the magnetic susceptibility and (b) pressure dependence of the peak of the magnetic susceptibility of MnSi according to [50]. Curves *1–11* in (a) respectively correspond to the pressures 0, 0.1, 0.15, 0.201, 0.268, 0.278, 0.303, 0.338, 0.379, 0.460, and 0.561 GPa. The symbols • and △ in (b) correspond to two different series of measurements. The height of the peak of the magnetic susceptibility can be seen to decrease sharply with the solidification of helium, which serves as the pressure-transmitting medium.

≈ 1.4 GPa, which suggested the possibility of observing quantum critical phenomena in MnSi.

Later, numerous studies appeared that confirmed this observation [24, 44–48]. One such study, [44], has played a special role in the subsequent investigations of MnSi (also see [45]). Based on measurements of the magnetic susceptibility, it was stated in [44] that a tricritical point with coordinates ≈ 12 K and ≈ 1.2 GPa exists on the phase transition line in MnSi. This conclusion was based on experimentally observed changes in the evolution of the magnetic susceptibility of MnSi in the vicinity of the phase transition [44, 45] with changing pressure (Fig. 15). The authors assumed that the degeneracy of the peak of magnetic susceptibility indicates a change in the order of the phase transition. A theory developed later stated the inevitability of a discontinuous character of the ferromagnetic phase transition at low temperatures [49].

New investigations of the magnetic phase diagram of MnSi, whose important feature was the use of helium as the pressure-transmitting medium, were performed in [50]. The results of these studies have demonstrated clearly (Fig. 16) that the magnetic susceptibility of MnSi is quite sensitive to nonhydrostatic stresses, which arise even in solid helium, and that the shape of the $\chi_{AC}(T)$ dependence cannot be used to decide on the character of the phase transition. On the other hand, the results of the investigations of the behavior of the MnSi resistivity and its temperature derivative $d\rho/dT$ in the vicinity of the phase transition at high pressures do not indicate a change in the character of the phase transition (Fig. 17). Therefore, taking the data given in Section 3.6 into account, we can assume that the magnetic phase transition in MnSi is a first-order phase transition in the entire range of the existence of the magnetic phase. Nonetheless, the behavior of the temperature derivative $d\rho/dT$ can hardly be interpreted unambiguously. As can be seen from Fig. 17a, the sharp peak of $d\rho/dT$ corresponding to the magnetic phase transition is broadened as the pressure increases, whereas the width of the anomaly (still of an unclear origin), which accompanies the phase transition, tends to zero (Fig. 17b). The last conclusion agrees with the results of measurements of the thermal expansion of MnSi [51]. We emphasize that the sharp peak in $\alpha(T)$ curves (see Figs 8, 9) cannot be resolved experimentally [51] because of the limitations of the method.

The behavior of the linear thermal expansion of MnSi $\Delta L/L$ in a wide range of pressures is shown in Fig. 18. It can be seen that the interval ΔT of the broad anomaly $\Delta L/L$ narrows strongly with increasing pressure, which gives the impression of a jump-like change in the volume ($\Delta L/L \sim 10^{-5}$) in MnSi with the phase transition at $T \to 0$ (Fig. 19). But the behavior of the extremely small volume jump ($\Delta L/L \sim 10^{-6}$) observed at atmospheric pressure remains unclear.

Returning to Fig. 17a, we note that the sharp peak and the broad anomaly appear to merge into a common maximum, leading, in the integral representation, to the picture shown in Figs 18 and 19. Nevertheless, the total situation is quite ambiguous and requires additional investigation.

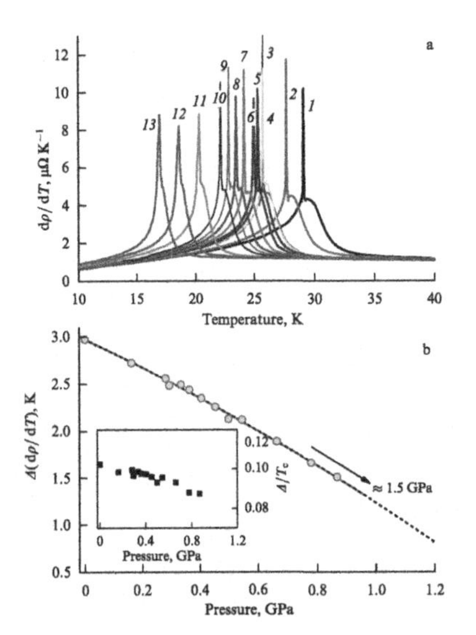

Figure 17. (a) Temperature dependence of the thermal resistivity coefficient $d\rho/dT$ of MnSi on T at different pressures in the region of the phase transition. (b) Pressure dependence of the width of the corresponding peaks $\Delta(d\rho/dT)$ measured at $d\rho/dT = 3$ μΩ cm K^{-1} [50]. Curves *1–13* in (a) correspond to the respective pressures 0, 0.163, 0.281, 0.294, 0.334, 0.363, 0.405, 0.453, 0.5, 0.55, 0.664, 0.779, and 0.868 GPa. The inset in (b) shows the pressure dependence of the ratio of the peak width to the phase transition temperature. The peak width is extrapolated to zero at the pressure corresponding to the quantum phase transition in MnSi at $T = 0$.

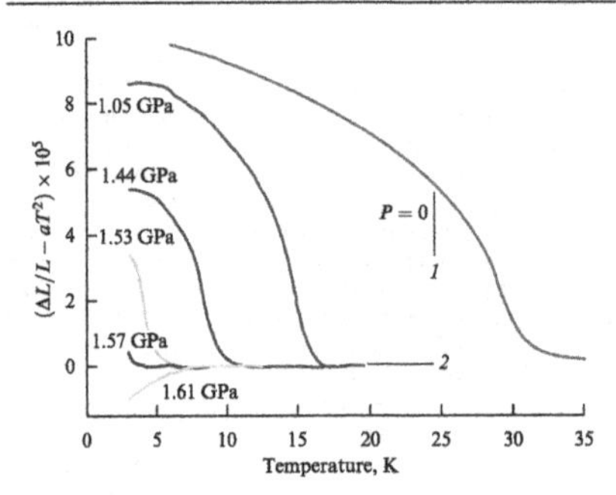

Figure 18. Thermal expansion of MnSi after the subtraction of the background component (aT^2) at various pressures: (*1*) according to [30]; (2) data from [51].

4.2 Non-Fermi-liquid behavior

The results mentioned in Section 4.1 appear to indicate that the quantum phase transition in MnSi at $T \to 0$ is a first-order phase transition, which generally excludes the development of quantum critical phenomena and related effects, such as a non-Fermi-liquid behavior of the resistivity and a logarithmic divergence of the heat capacity.

As was shown in [3, 4], the resistivity of MnSi in the paramagnetic phase at $P > P_c$ ($\approx$ 15 kbar) and $T < 5$ K is described by a power law $\Delta\rho \propto T^n$ with the exponent $n = 1.5$ (we recall that $n = 2$ for the Fermi liquid and $n = 1$ in the quantum critical region) (Fig. 12). It is surprising that the region of the non-Fermi-liquid behavior extends at least to the pressure of 48 kbar ($\approx 3P_c$). With a further increase in pressure, the exponent n increases to 1.7 [53].

However, the data in [53] should be considered with a certain amount of skepticism, because the conditions of the corresponding measurements can hardly be called hydrostatic. Nevertheless, the very fact of the non-Fermi-liquid behavior of the temperature dependence of the resistivity of MnSi in a significant range of pressures $P > P_c$ is beyond dispute (Fig. 20).

The character of the manifestations of the non-Fermi-liquid behavior in MnSi suggests that it can have no relation to the quantum criticality, which poses a question on the origin of this phenomenon. In this connection, we should keep in mind that the $\sim T^{3/2}$ dependence of the electrical resistivity has been found in many amorphous metals and spin glasses [54]. The theory of resistivity of MnSi with the scattering of electrons on skyrmions taken into account to explain the $T^{3/2}$ law was suggested in [55] (for skyrmions, see Appendix I).

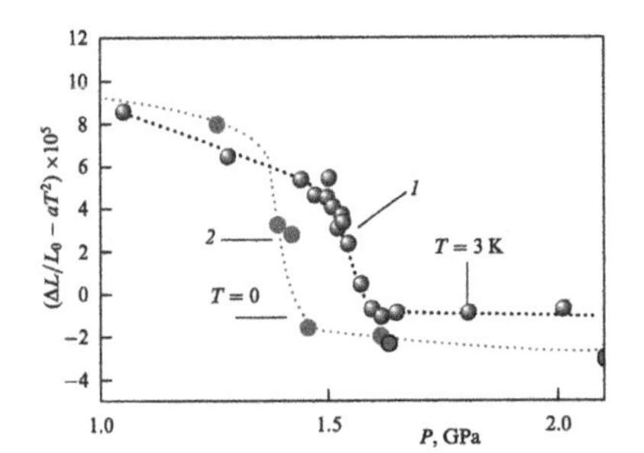

Figure 19. Relative variations of (*1*) the length of an MnSi sample [51] and (*2*) the lattice parameter of MnSi [52] at low temperatures and high pressures.

4.3 'Partial' helical order in the non-Fermi-liquid phase of MnSi

Figure 21 presents the results of small-angle neutron scattering (SANS) investigations of the magnetic superstructure in MnSi at ambient and high pressures according to [5]. It follows from this figure that the Bragg reflections with the wave vector **q** ($q = 0.037$ Å^{-1}) directed along [111], which characterize the magnetic spiral at ambient pressure, do not disappear completely in the region of the non-Fermi-liquid paramagnetic phase. The measurements at $P > P_c$ and $T < T_0$ (Fig. 21) reveal the existence of reflections corresponding to the magnetic spiral with the wave vector **q** ($q \approx 0.043$ Å^{-1}) continuously distributed over the angular space, with a maximum in the [110] direction. According to [5], the magnetic phase transition at P_c is therefore a reorientation phase transition and is determined by the

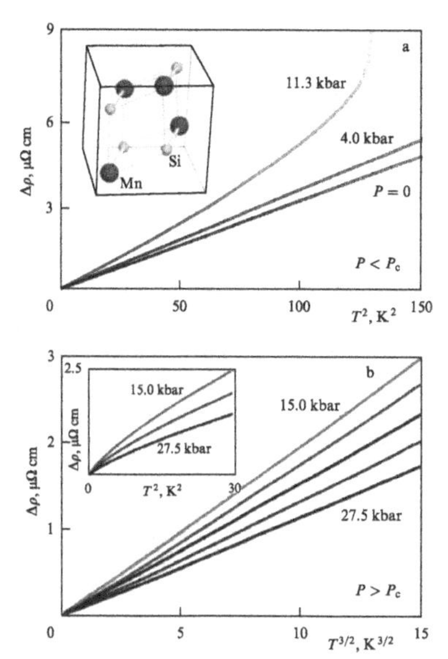

Figure 20. Electrical resistivity of MnSi as a function of temperature at high pressures [4] ($\Delta\rho = \rho - \rho_{T=0}$; P_c is the pressure of the quantum phase transition at $T = 0$).

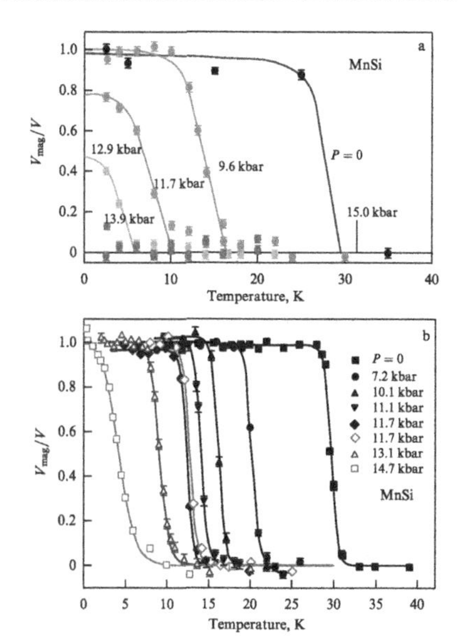

Figure 22. Magnetic volume fraction V_{mag}/V in MnSi as a function of temperature at high pressures: (a) data from [56] and (b) data from [58].

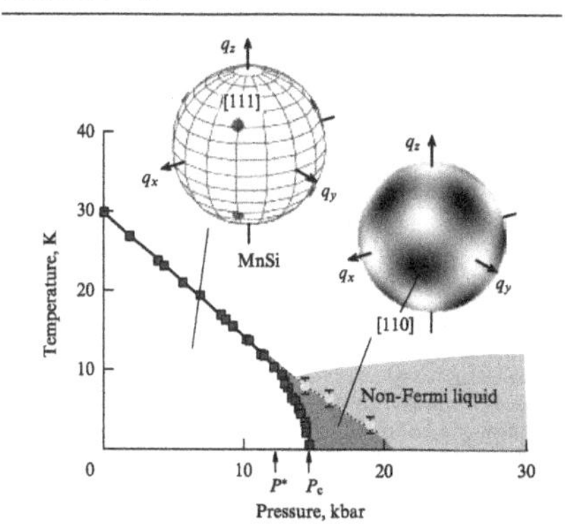

Figure 21. Small-angle neutron scattering in MnSi at ambient and high pressures (schematic) [5]. Square symbols correspond to the magnetic phase transition temperature; circles correspond to the temperature of the crossover between the paramagnetic phase and the state of partial ordering; P^* is the pressure corresponding to a hypothetical tricritical point; P_c is the pressure of the magnetic phase transition at P^*.

smallest of the three characteristic energies (see Section 2) responsible for the direction of the magnetic spiral in MnSi.

On the other hand, we warn readers against a noncritical perception of the above results. The use of fluorinert as a pressure-transmitting medium in neutron-diffraction investigations at high pressures is not optimal from the standpoint of conditions of hydrostaticity and sometimes leads to wrong conclusions.

4.4 Phase separation in MnSi at high pressures

The phase separation observed in a number of magnetic systems is by no means a universal indicator of a first-order phase transition, whereas such a transition is necessarily accompanied by phase separation.

In this connection, it is interesting to investigate the phase separation in MnSi at high pressures using muon spin spectroscopy (μSR) [56, 58]. It has been stated that the content of the magnetic fraction decreases rapidly starting with a pressure of 11.7 kbar, which means the existence of a phase separation outside the first-order phase transition region (Fig. 22a). This fact was given special importance, which we do not fully appreciate [56]. The phase inhomogeneity in MnSi at high pressures was also noted in [57].

However, new μSR investigations of MnSi [58] indicate an almost 100% amount of the magnetic fraction at pressures up to the value corresponding to the phase transition (Fig. 22b). The origin of the contradiction appears to lie in the violation of the hydrostaticity condition in investigations at high pressures (see the discussion in [58]).

5. Conclusions

To summarize, two problems that require special attention should be noted, in our opinion.

(1) As is emphasized in this review, the heat capacity, the thermal expansion coefficient, the thermal resistivity coefficient, elastic moduli, and the ultrasound attenuation coefficient during the phase transition in MnSi exhibit flat maxima (or a minimum in the case of the thermal expansion coefficient) at a temperature that is somewhat higher than T_c (see Section 3). In some cases, this feature is so strongly pronounced (see, e.g., the behavior of elastic moduli in Fig. 10) as to suggest the occurrence of some global transformation in the spin subsystem of MnSi, whose nature is yet to be clarified. In view of this impressive feature, the phase transition at $T \approx 29$ K seems only an insignificant event. This situation is excellently illustrated by the behavior of the entropy (see Fig. 14).

The clarification of the nature of this transformation and of its interrelation with the phase transition appears to be one of the primary problems to be solved.

(2) It follows from the results of measurements of thermal expansion of MnSi at high pressures (see Fig. 18) that the wide large anomaly corresponding to the flat minimum of the thermal expansion coefficient narrows strongly as the pressure increases, which possibly suggests a jump-like volume change in MnSi in the phase transition in the limit as $T \to 0$. This observation agrees with the data in Fig. 17, where the sharp peak of $d\rho/dT$ corresponding to the magnetic phase transition is broadened with increasing pressure, whereas the width of the flat anomaly of unclear origin, which accompanies the phase transition, tends to zero (Fig. 17b). Apparently, the sharp peak and the wide anomaly merge into a common maximum, producing a kind of a quantum tricritical point, as is shown in the hypothetical phase diagram in Fig. 23.

The vicinity of the quantum phase transition in MnSi at the pressure $P_c \approx 15$ kbar is therefore a kind of 'treasure island,' whose map is fully packed with labels for 'skyrmions,' 'partial spin order,' 'non-Fermi-liquid phase,' and 'quantum tricritical point.' However, no complete certainty exists that the labels are located correctly or that their names correspond to reality. It is also unclear whether there exist paths that connect these labels to a common system.

Therefore, detailed investigations of the phase diagram of MnSi at high pressures and low temperatures under truly hydrostatic conditions seem to be a necessary step in solving the above problems. However, such experiments seem to be hardly feasible at present in view of the extremely high sensitivity of MnSi to nonhydrostatic stresses. New approaches should be developed in the field of low-temperature techniques of high pressures.

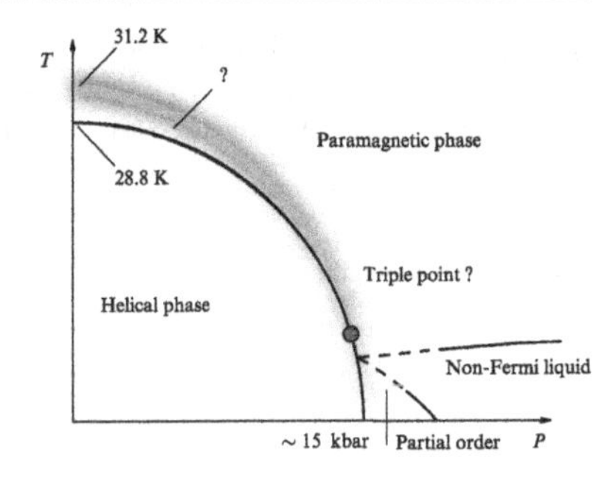

Figure 23. Hypothetical phase diagram of MnSi.

6. Appendices

Appendix I: Nature of the A phase and skyrmions

The wave vector of the magnetic spiral in MnSi in a zero magnetic field is known to be oriented in the direction of the spatial diagonal of the cubic unit cell [111]. At $T < T_c$, four domains are formed in MnSi, corresponding to four equivalent [111] directions. At low temperatures, the application of a magnetic field $B \approx 0.1$ T leads to the monodomainization of the sample, which acquires a conic magnetic structure and eventually passes into a field-induced ferromagnetic state at $B \approx 0.6$ T [28]. A somewhat different situation arises at temperatures close to T_c. In this case, as was established by the investigations of ultrasound, magnetization, and so on [29], a so-called A phase exists in magnetic fields $\approx 0.120-0.200$ T at temperatures from T_c to ≈ 27 K, whose spin structure is still the subject of debate. Initially, it the A phase was assumed to be paramagnetic [29], but experiments on neutron scattering revealed the existence of a magnetic order [59, 60].

The first ideas concerning the magnetic structure of the A phase, which were based on neutron-diffraction studies, amounted to the assumption that the wave vector of the magnetic spiral in the A phase becomes oriented perpendicular to the direction of the magnetic field [61]. But the recent neutron-diffraction studies performed in a special geometry indicate, according to the authors of [9], in favor of the skyrmion-related nature of the magnetic order in the A phase. In fact, the authors of [9] speak of the observation of a two-dimensional lattice of skyrmion vortices, which can be represented as a superposition of three spirals arranged at an angle of 120° and located in a plane perpendicular to the magnetic field vector [8, 62]. The corresponding measurements performed in [63] using solid solutions $Mn_{1-x}Fe_xSi$, $Mn_{1-x}Co_xSi$, and $Fe_{1-x}Co_xSi$, which crystallize, just as MnSi, in the B20-type structure, led these authors to analogous conclusions. Finally, the direct observations of a skyrmion lattice in $Fe_{1-x}Co_xSi$ and FeGe performed using Lorentz electron microscopy [64, 65] should be mentioned. Nevertheless, as was noted in [65], the structural data do not allow clarifying whether the spin structure observed in the A phase is a true skyrmion crystal. It has been claimed that this problem can be solved by the investigation of the so-called topological Hall effect [66].

We note that the ideas on the formation of skyrmions in spin systems with a chiral spin–orbital interaction were first suggested in [67, 68].

Appendix II: Heat capacity and entropy of MnSi at $T > T_c$

Various contributions to the heat capacity of MnSi were analyzed based on the experimental and calculated data that characterize the phonon spectrum of MnSi [33]. Figure 24 illustrates the behavior of the heat capacity of MnSi in a zero magnetic field and in the field $B = 4$ T. We can see that the magnetic field suppresses the phase transition, as was shown in Section 2.2 (see Figs 7 and 9). On the other hand, the heat capacity of MnSi in the magnetic field $B = 4$ T is higher than

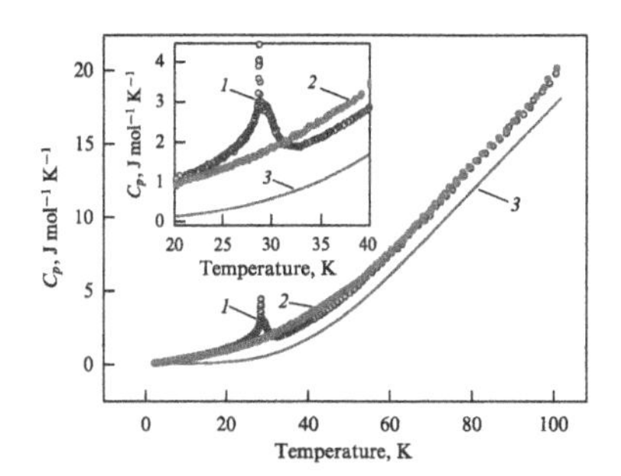

Figure 24. Heat capacity of MnSi at (1) $B = 0$ and (2) $B = 4$ T and (3) the calculated phonon contribution to heat capacity of MnSi.

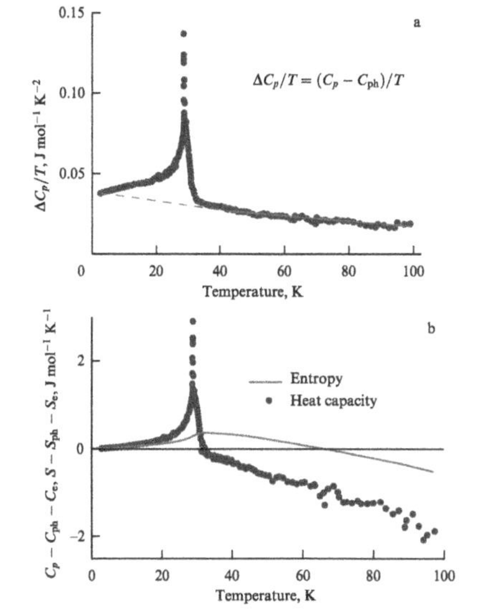

Figure 25. Temperature dependence of (a) $\Delta C_p/T = (C_p - C_{\rm ph})/T$ ratio and (b) the 'magnetic' heat capacity and entropy for MnSi. $C_{\rm pg}$ is the phonon contribution.

that at $B = 0$, which seems to be somewhat strange, because the magnetic field suppresses spin fluctuations.

Figure 25 displays the temperature dependence of the ratio $\Delta C_p/T = (C_p - C_{\rm ph})/T$, where $C_{\rm ph}$ is the phonon part of the heat capacity. It can be seen from the figure that the linear (electron-related) term γ in the heat capacity of MnSi has the same value at both $T < T_c$ and $T > T_c$. By subtracting the electron and phonon components from the total heat capacity, the authors of [33] came to a conclusion on the existence of negative contributions to the heat capacity and entropy of MnSi at $T > T_c$ and $B = 0$ (Fig. 25b). We recall that such a behavior of the heat capacity was predicted many years ago as a result of the existence of nonlinear effects in fluctuating spin systems [69]. The authors of [33] assume that the decrease in the partial entropy means the appearance of some ordering in the spin subsystem. However, it is impossible at present to decide whether this situation is related to the formation of skyrmion-like vortex structures or it merely reflects the tendency to spin localization.

Acknowledgments. This work was supported in part by the Russian Foundation for Basic Research (project no. 09-02-00336).

References

1. Borén B *Arkiv Kemi Min. Geol.* **11A** 1 (1933)
2. Thompson J D, Fisk Z, Lonzarich G G *Physica B* **161** 317 (1989)
3. Pfleiderer C, Julian S R, Lonzarich G G *Nature* **414** 427 (2001)
4. Doiron-Leyraud N et al. *Nature* **425** 595 (2003)
5. Pfleiderer C et al. *Nature* **427** 227 (2004)
6. Rößler U K, Bogdanov A N, Pfleiderer C *Nature* **442** 797 (2006)
7. Binz B, Vishwanath A, Aji V *Phys. Rev. Lett.* **96** 207202 (2006)
8. Tewari S, Belitz D, Kirkpatrick T R *Phys. Rev. Lett.* **96** 047207 (2006)
9. Mühlbauer S et al. *Science* **323** 915 (2009)
10. Tanaka M et al. *J. Phys. Soc. Jpn.* **54** 2970 (1985)
11. Ishida M et al. *J. Phys. Soc. Jpn.* **54** 2975 (1985)
12. Williams H J et al. *J. Appl. Phys.* **37** 1256 (1966)
13. Ishikawa Y et al. *Solid State Commun.* **19** 525 (1976)
14. Dzyaloshinsky I *J. Phys. Chem. Solids* **4** 241 (1958)
15. Moriya T *Phys. Rev.* **120** 91 (1960)
16. Kataoka M et al. *J. Phys. Soc. Jpn.* **53** 3624 (1984)
17. Grigoriev S V et al. *Phys. Rev. B* **81** 012408 (2010)
18. Petrova A E et al. *Phys. Rev. B* **82** 155124 (2010)
19. Jeong T, Pickett W E *Phys. Rev. B* **70** 075114 (2004)
20. Pan Z J, Zhang L T, Wu J S *J. Appl. Phys.* **101** 033715 (2007)
21. Mattheiss L F, Hamann D R *Phys. Rev. B* **47** 13114 (1993)
22. Wernick J H, Wertheim G K, Sherwood R C *Mat. Res. Bull.* **7** 1431 (1972)
23. Fawcett E, Maita J P, Wernick J H *Int. J. Magnetism* **1** 29 (1970)
24. Thessieu C et al. *Solid State Commun.* **95** 707 (1995)
25. Petrova A E et al. *Phys. Rev. B* **74** 092401 (2006)
26. Pfleiderer C *J. Magn. Magn. Mater.* **226–230** 23 (2001)
27. Stishov S M et al. *Phys. Rev. B* **76** 052405 (2007)
28. Ishikawa Y, Komatsubara T, Bloch D *Physica B* **86–88** 401 (1977)
29. Ishikawa Y, Arai M *J. Phys. Soc. Jpn.* **53** 2726 (1984)
30. Stishov S M et al. *J. Phys. Condens. Matter* **20** 235222 (2008)
31. Lamago D et al. *Physica B* **385–386** 385 (2006)
32. Bak P, Jensen M H *J. Phys. C* **13** L881 (1980)
33. Stishov S M et al. *Phys. Rev. Lett.* **105** 236403 (2010)
34. Belitz D, Kirkpatrick T R, Rosch A *Phys. Rev. B* **73** 054431 (2006)
35. Janoschek M et al. *Phys. Rev. B* **81** 214436 (2010)
36. Matsunaga M, Ishikawa Y, Nakajima T *J. Phys. Soc. Jpn.* **51** 1153 (1982)
37. Petrova A E, Stishov S M *J. Phys. Condens. Matter* **21** 196001 (2009)
38. Petrova A E, Stishov S M *Prib. Tekh. Eksp.* (4) 173 (2009) [*Instrum. Exp. Tech.* **52** 609 (2009)]
39. Kusaka S et al. *Solid State Commun.* **20** 925 (1976)
40. Nakanishi O et al. *Solid State Commun.* **35** 995 (1980)
41. Diep H T *Phys. Rev. B* **39** 397 (1989)
42. Plumer M L, Mailhot A *Phys. Rev. B* **50** 16113 (1994)
43. Pappas C et al. *Phys. Rev. Lett.* **102** 197202 (2009)
44. Pfleiderer C, McMullan G J, Lonzarich G G *Physica B* **206–207** 847 (1995)
45. Pfleiderer C et al. *Phys. Rev. B* **55** 8330 (1997)
46. Pfleiderer C et al. *Physica B* **230–232** 576 (1997)
47. Thessieu C, Pfleiderer C, Flouquet J *Physica B* **239** 67 (1997)
48. Thessieu C, Kitaoka Y, Asayama K *Physica B* **259–261** 847 (1999)
49. Belitz D, Kirkpatrick T R, Vojta T *Phys. Rev. Lett.* **82** 4707 (1999)
50. Petrova A E et al. *Phys. Rev. B* **79** 100401(R) (2009)
51. Miyake A et al. *J. Phys. Soc. Jpn.* **78** 044703 (2009)
52. Pfleiderer C et al. *Science* **316** 1871 (2007)
53. Pedrazzini P et al. *Physica B* **378–380** 165 (2006)

54. Ford P J, Mydosh J A *Phys. Rev. B* **14** 2057 (1976)
55. Kirkpatrick T R, Belitz D *Phys. Rev. Lett.* **104** 256404 (2010)
56. Uemura Y J et al. *Nature Phys.* **3** 29 (2007)
57. Yu W et al. *Phys. Rev. Lett.* **92** 086403 (2004)
58. Andreica D et al. *Phys. Rev. B* **81** 060412(R) (2010)
59. Lebech B, in *Recent Advances in Magnetism of Transition Metal Compounds* (Eds A Kotani, N Suzuki) (Singapore: World Scientific, 1993) p. 167
60. Lebech B et al. *J. Magn. Magn. Mater.* **140–144** 119 (1995)
61. Grigoriev S V et al. *Phys. Rev. B* **73** 224440 (2006)
62. Binz B, Vishwanath A *Phys. Rev. B* **74** 214408 (2006)
63. Münzer W et al. *Phys. Rev. B* **81** 041203(R) (2010)
64. Yu X Z et al. *Nature* **465** 901 (2010)
65. Yu X Z et al. *Nature Mater.* **10** 106 (2011)
66. Neubauer A et al. *Phys. Rev. Lett.* **102** 186602 (2009)
67. Bogdanov A N, Yablonskii D A *Zh. Eksp. Teor. Fiz.* **95** 178 (1989) [*Sov. Phys. JETP* **68** 101 (1989)]
68. Bogdanov A, Hubert A *J. Magn. Magn. Mater.* **138** 255 (1994)
69. Murata K, Doniach S *Phys. Rev. Lett.* **29** 285 (1972)

A.8.2. Helical itinerant MnSi magnet: magnetic phase transition

Physics – Uspekhi **60** (12) 1268 – 1276 (2017)

PHYSICS OF OUR DAYS

PACS numbers: **62.50.–p**, 75.30.Kz, 75.40.Cx, 77.80.B–

Helical itinerant MnSi magnet: magnetic phase transition

S M Stishov, A E Petrova

DOI: https://doi.org/10.3367/UFNe.2017.03.038110

Contents

Abstract. New studies of the phase transition and phase diagram of the chiral MnSi magnet are reported. New results are obtained in the course of analysis of experimental data on heat capacity, thermal expansion, elastic properties, electrical resistance, neutron scattering, and theoretical modeling.

Keywords: helical magnet, magnetic phase diagram, phase transitions

1. Introduction

MnSi — a helical itinerant magnet — is crystallized in the structure type B20 belonging to the space group $P2_13$, which has no center of symmetry. Although it has been studied intensively for several decades, this magnetic material still remains at the center of attention of many researchers. This circumstance is due to a whole series of reasons, some of which are as follows.

(1) MnSi is an example of a substance with a helical magnetic structure caused by the Dzyaloshinskii–Moriya interaction [1].

(2) Studies of the physical properties of MnSi at high pressures have revealed a number of intriguing features, such as a quantum phase transition [1, 2], non-Fermi-liquid behavior [3, 4], and 'partial' helical order [5], which still await further study.

(3) The so-called phase A, appearing in MnSi in a magnetic field, which was identified as a skyrmion crystal [1, 6], has proven to be extremely sensitive to various kinds of actions and is considered a promising material for spintronics [7].

S M Stishov, A E Petrova Vereshchagin Institute of High Pressure Physics, Russian Academy of Sciences,
Kaluzhskoe shosse 14, 108840 Troitsk, Moscow, Russian Federation
E-mail: sergei@hppi.troitsk.ru

Received 6 March 2017, revised 23 March 2017
Uspekhi Fizicheskikh Nauk **187** (12) 1365 – 1374 (2017)
DOI: https://doi.org/10.3367/UFNr.2017.03.038110
Translated by S N Gorin; edited by A Radzig

(4) The phase transition in MnSi at a temperature of 29 K and normal pressure from the paramagnetic state into the helical state has finally been acknowledged as a first-order phase transition [8], which casts doubts upon early assertions about the shape of the phase diagram of MnSi at high pressures [9].

(5) Finally, which is of great importance, MnSi constitutes a simple binary compound synthesized from elements that allow deep purification. MnSi possesses a comparatively low temperature of congruent melting (~ 1500 K), which facilitates growing large crystals. To date, large and sufficiently perfect MnSi single crystals of high purity have been grown in a number of laboratories, and thus have become accessible for diverse physical studies.

This article comprises a survey of new data obtained in the course of studies on heat capacity, thermal expansion, elastic properties, electrical resistance, and neutron scattering in MnSi, in particular, at high pressures and in strong magnetic fields.

2. Magnetic phase transition in MnSi

Let us first examine the magnetic phase transition in MnSi at atmospheric pressure in a zero magnetic field. The analysis of a phase transition in any system, as a rule, begins from the mean-field approximation, which disregards fluctuations. The most popular approximation is the phenomenological Landau model based on symmetry considerations, in which the expansion of the thermodynamic potential Φ in powers of the order parameter η is used. In the simplest case, the Landau expansion is written out as follows:

$$\Phi(P, T, \eta) = \Phi_0 + A\eta^2 + C\eta^3 + B\eta^4 + \dots , \qquad (1)$$

where the coefficients A, B, and C are functions of pressure P and temperature T. In view of the symmetry relative to time reversal, $C \equiv 0$ in the case of magnetic phase transitions, and relationship (1) takes on the following form:

$$\Phi(P, T, \eta) = \Phi_0 + A\eta^2 + B\eta^4 + \dots \qquad (2)$$

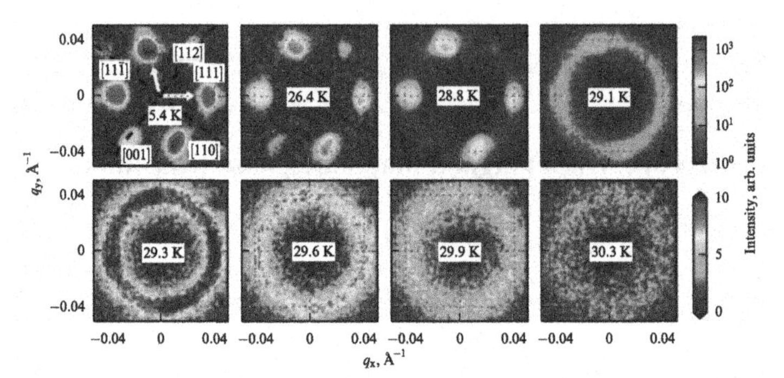

Figure 1. (Color online.) Helicomagnetic phase transition in MnSi at 29 K according to data from neutron experiments [16]; q_x and q_y are the coordinates in the momentum space. It can be seen that, during a phase transition in the temperature interval of 28.8–29.1 K, the discrete Bragg peaks corresponding to the helical magnetic order, are replaced by diffuse magnetic scattering, which is concentrated on the surface of a sphere in the reciprocal space. These effects reflect the existence of helical magnetic fluctuations in the immediate proximity to the temperature of the phase transition.

At the point of the second-order phase transition, one has $A = a(T - T_c) = 0$ and the coefficient $B > 0$. For $B < 0$, a first-order phase transition takes place in the system. In this case, a sixth-order term $D\eta^6$ with $D > 0$ should be added to expansion (2) to stabilize the system. The available experimental data indicate a certain lability of the coefficient B in different systems; as a result, the phase transition can alter its nature with a change in the ambient conditions with the appearance of the so-called tricritical point [10].

There are internal reasons for which some phase transitions are first-order in spite of the absence of symmetry limitations. First, there are striction effects, which can arise in real compressible lattices [11, 12]. Competition among two or more order parameters caused by different interactions, if accompanied by strong fluctuations, can also lead to a first-order phase transition. In essence, these factors renormalize the coefficient B and even lead to the appearance of nonzero third-order terms (see the Halperin–Lubensky–Ma effect [13]). No one should be embarrassed by the fact that we are discussing here the evolution of the Landau expansion coefficients, which are the mean-field parameters, together with the fluctuation effects, which seemingly violate the mean-field picture of the phase transition. In reality, the fluctuations are concentrated in a relatively narrow temperature range determined by the Levanyuk–Ginzburg criterion [10], and beyond this range the mean-field picture remains valid; therefore, we can consider fluctuations as a factor that can affect both the sign and the value of the coefficients of the Landau expansion.

Further on, turning to the subject of our study it should be noted that the fluctuational nature of the first-order phase transition in MnSi was indicated for the first time by Bak and Jensen [14], who carried out relevant calculations, apparently influenced by the work of Hansen [15], who revealed an abrupt change in the intensity of the line of superlattice scattering at the point of the phase transition. By the way, Hansen's work was not published in the contemporary scientific literature and remained unknown, until it was mentioned as a reference cited in paper [16].

At present, the fluctuational model of the phase transition in MnSi, based on Brazovskii's theory [17], is becoming popular; here, the role of fluctuations in systems with a fluctuation spectrum possessing a low absolute minimum with a nontrivial value of the moment is emphasized. The authors of Ref. [16] have analyzed the experimental data on small-angle neutron scattering in the vicinity of the phase transition in MnSi and developed a theory which, in their opinion, describes well the totality of the experimental data (Fig. 1).

It is necessary to recall that an important feature of the magnetic phase transition in MnSi is the presence of a secondary gently sloping maximum or a shoulder in the curves of the heat capacity, thermal expansion, thermal resistance, and sound absorption, as is evident, for example, from Fig. 2. The shoulder is located in the existence domain boundaries of the paramagnetic phase but is characterized by powerful chiral fluctuations, as follows from the data on neutron scattering [18]. According to the authors of paper [16], they obtained a successful description of this feature of the phase transition in MnSi, although their graphic proof of the existence of a shoulder (see Fig. 2) is by no means convincing.

3. Vollhardt 'invariant'

Let us say several words about the so-called Vollhardt invariant. In 1997, D Vollhardt revealed that the curves of the heat capacity of a number of highly correlated systems measured at different values of the thermodynamic parameters (for example, pressure, magnetic field, etc.) intersect at almost the same temperature [19]. This effect is distinctly visible in the curves of heat capacity, thermal expansion, elastic moduli, and sound absorption (Fig. 3) [20]. The very term 'Vollhardt invariant' was apparently first used in Ref. [21], where the data on the heat capacity of MnSi, Mn(Co)Si, and Mn(Fe)Si were analyzed.

The presence of this invariant is connected with the existence of a specific characteristic energy determining the population of helical magnetic fluctuations, which naturally indicates the occurrence of the Dzyaloshinskii–Moriya interaction [21]. This conclusion is supported in Ref. [16], where it is asserted that the appearance of the Vollhardt

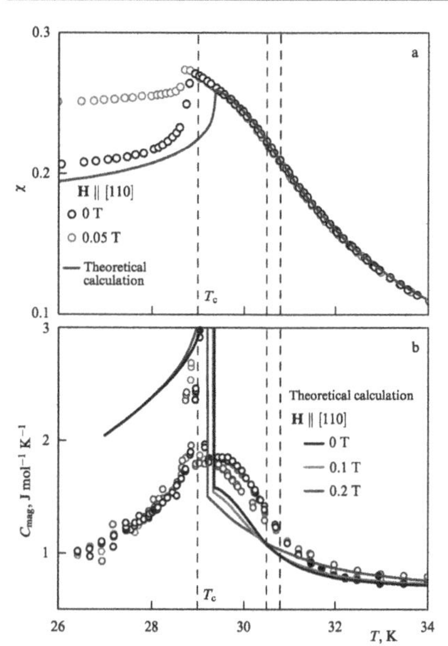

Figure 2. (Color online.) Comparison of the results of theoretical calculations based on the Brazovskii model [16] and experimental data, which characterize (a) the magnetic susceptibility χ, and (b) the heat capacity C_{mag} of MnSi according to the data from Ref. [16]. The solid curves correspond to the theoretical calculations; the circles show experimental data.

invariant directly follows from the Brazovskii model [17]. A simulation of the situation with the aid of Gaussian functions performed in Ref. [20] shows that the appearance of a pseudoinvariant temperature should always be expected when a magnetic or some other field broadens the appropriate maxima and decreases their amplitudes in such a manner that their integral values remain unaltered (Fig. 4).

Thus, in Ref. [20] we reached the conclusion that the 'point of intersection' or the Vollhardt invariant cannot serve as an indicator of the existence of some specific energy, and the very 'invariant' in general simply represents an approximate point of the intersection of the corresponding curves rather than an invariant, as can be seen from Fig. 3. The very maxima or minima, illustrated in Fig. 5, are identified in Ref. [20] as smeared phase transitions.

4. Phase transition in MnSi according to simulations by the Monte Carlo method

Here, we should pay special attention to paper [22], which is devoted to an analysis of the properties of a three-dimensional lattice system of spins with the aid of the classical Monte Carlo method. Together with the ferromagnetic exchange interaction (J), the authors of Ref. [22] take into account the anisotropic Dzyaloshinskii–Moriya interaction (D). Figure 6, borrowed from Ref. [22], demonstrates an excellent agreement of the results of simulations with the

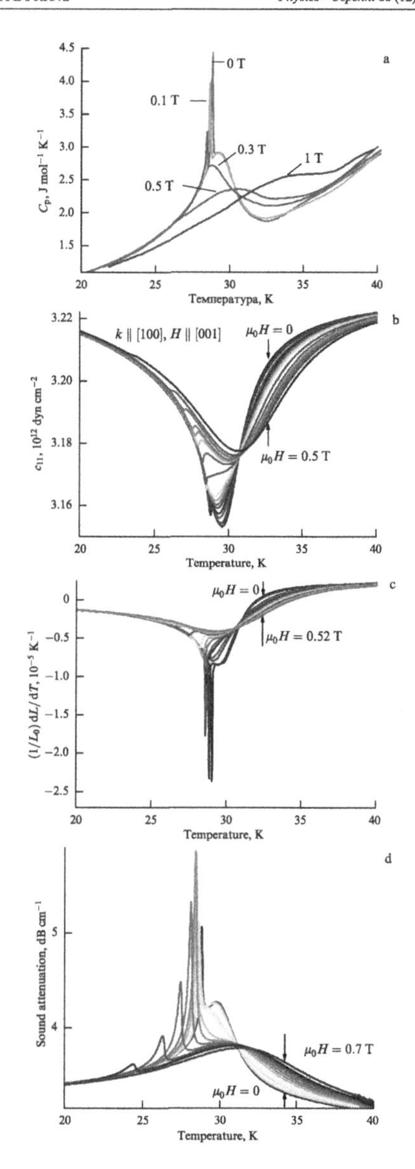

Figure 3. (Color online.) (a) Heat capacity, (b) elastic modulus c_{11}, (c) coefficient of thermal expansion, and (d) coefficient of sound absorption as functions of temperature and magnetic field in a phase transition in MnSi [20]; μ_0 is the permeability of vacuum.

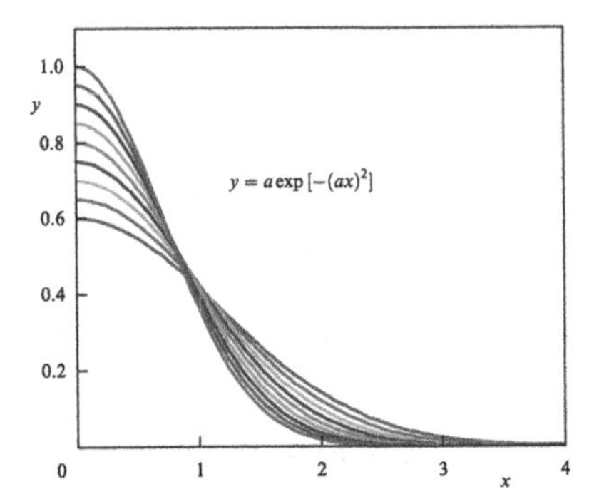

Figure 4. (Color online.) Simulation of the effect of the Volkhardt intersection with the aid of a Gaussian function and a variation of the width a [20].

experimental data for MnSi in a zero magnetic field. In Ref. [23], a study of the same system of spins was carried out by the same numerical method but with a variable ratio between the exchange interaction and the Dzyaloshinskii–Moriya interaction (J/D). The use of the J/D ratio should not be misleading, since the terms containing J and D enter into the Hamiltonian as summands, and the zeroing of the corresponding terms does not lead to divergence.

As it turns out, the shoulder in the curve of the heat capacity of the model system of spins appears as a result of a perturbation of the ferromagnetic second-order phase transition by helical fluctuations caused by the Dzyaloshinskii–Moriya interaction. With an increase in the contribution of this interaction, a first-order phase transition occurs in the system, and at $J/D \approx 1$ the behavior of the heat capacity of the system becomes similar to that experimentally observed in the cases of MnSi and Cu_2OSeO_3 (see Section 6): a sharp peak appears, which corresponds to a first-order transition, as does a gently sloping maximum or a shoulder at temperatures that somewhat exceed the temperature of the phase transition (Fig. 3).

As follows from the calculations carried out, the observed maxima in the temperature dependences of the physical quantities (Fig. 7) are connected with the 'smeared' second-order phase transition, which is in complete agreement with the conclusion made in Ref. [20]. However, it was assumed in Ref. [20] that the smearing is connected with a natural imperfection of the MnSi crystal, whereas it follows from the calculations performed by the Monte Carlo method that the degradation of the ferromagnetic second-order phase transition is a result of helical fluctuations.

Thus, the region of the maximum possesses a complex structure, which corresponds to the interaction of two fluctuating order parameters; as a result, the system cannot pass into the ordered state continuously, but does so jumpwise, via a first-order transition.

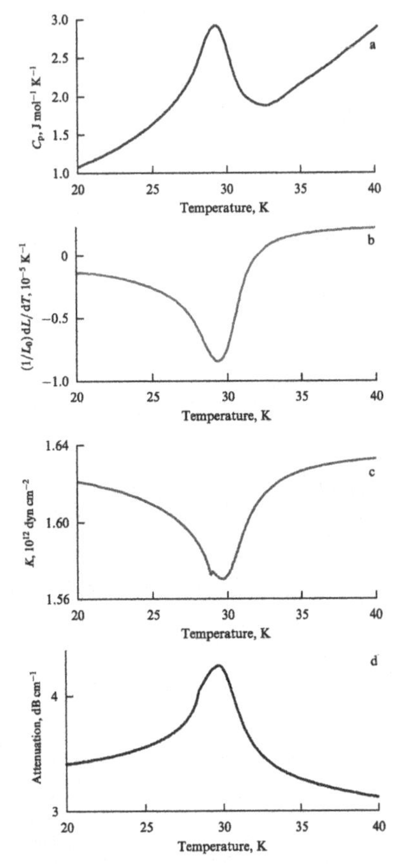

Figure 5. (a) Heat capacity, (b) coefficient of thermal expansion, (c) bulk modulus K, and (d) coefficient of sound absorption as functions of temperature in a phase transition in MnSi [20]. The sharp peaks caused by the first-order phase transition are removed to more distinctly show the anomalies, which, apparently, are characteristic of the smeared phase transition.

5. Phase transition in MnSi in a magnetic field

Let us now turn to studies of the phase transition in MnSi in a magnetic field. In Ref. [24], detailed measurements of the heat capacity of MnSi in magnetic fields were carried out, which revealed a nontrivial behavior of the heat capacity in the vicinity of the phase transition (Fig. 8). The authors of Ref. [24] assume that the results presented in Fig. 8 indicate the existence of a tricritical point at a temperature of 28.5 K and a magnetic field of 340 mT (Fig. 9). Simultaneously, the authors emphasize that the observed anomalies in the behavior of heat capacity at the boundaries of the skyrmion phase indicate their thermodynamic nature.

The results of ultrasonic studies of the phase diagram of MnSi in a magnetic field are presented in Figs 10 and 11 [25]. Notice that the studied MnSi sample had the shape of a disk, which affected the value of the demagnetization factor at

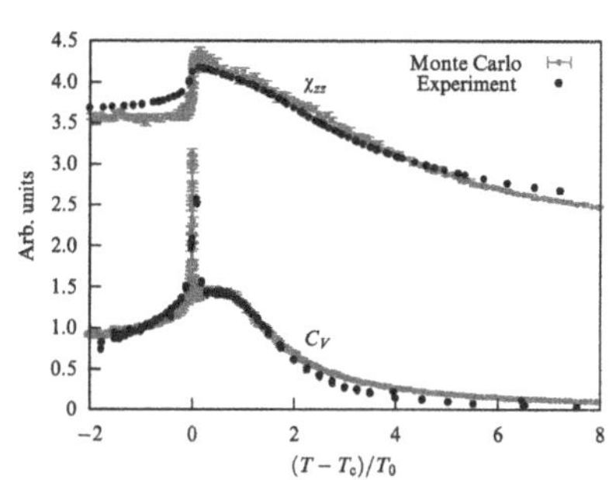

Figure 6. (Color online.) Heat capacity and the magnetic susceptibility in the phase transition in a three-dimensional system of Heisenberg spins with the Dzyaloshinskii–Moriya interaction according to data calculated by the Monte Carlo method [22]. Experimental data for MnSi were borrowed from Ref. [16].

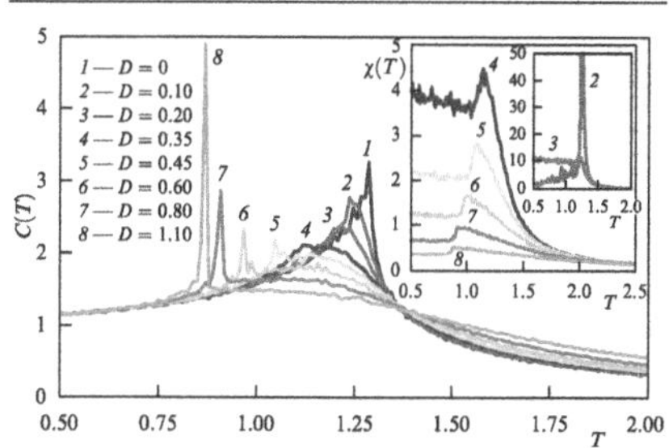

Figure 7. (Color online.) Behavior of the heat capacity at the phase transition in a three-dimensional system of Heisenberg spins with the Dzyaloshinskii–Moriya interaction with a variation in the Dzyaloshinskii–Moriya coupling constant D and the exchange interaction constant $J = 1$ according to data of simulations by the Monte Carlo method [23]. Temperature is given in units of J; the heat capacity, in dimensionless units. In the insets, the magnetic susceptibility $\chi(T)$ is shown at different values of D.

various orientations of the sample in the magnetic field. As a result, the magnetic scale of the corresponding dependences proves to be somewhat different (see Figs 10 and 11). Notice also that the higher the uniformity of the magnetic field in the sample, the less the domain of existence of the phase of the

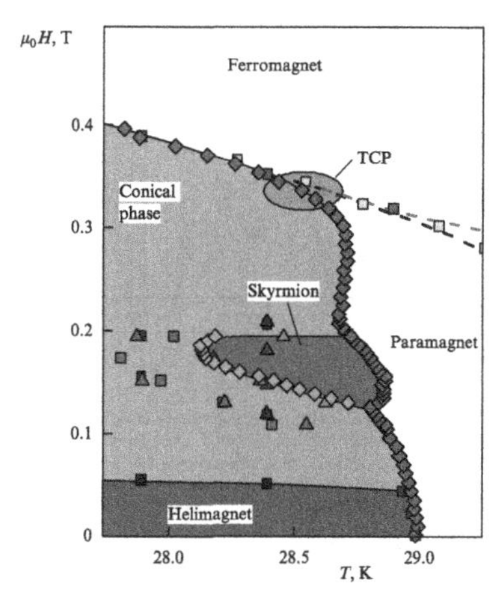

Figure 9. (Color online.) Magnetic phase diagram of MnSi. The oval contour outlines the region of the location of the supposed tricritical point (TCP) [24].

skyrmion crystal (in the $\mathbf{k} \perp \mathbf{H}$ configuration, the magnetic field is directed parallel to the plane of the disk and is distributed more uniformly than in the case of $\mathbf{k} \parallel \mathbf{H}$).

As can be seen from Fig. 10, an abrupt change in the elastic moduli c_{11} and c_{33} in the magnetic phase transition in MnSi upon the imposition of a magnetic field first diminishes rapidly, reaching practically zero values in the domain of existence of the skyrmion phase, then grows in the region of magnetic fields and temperatures indicated in Ref. [24] as the tricritical coordinates, and further decreases to negligibly small values. All this is illustrated in Fig. 12, which demonstrates the dependence of the jumps of the moduli and the amplitude of the attenuation factor on the magnetic field at different orientations. Let us emphasize that at the tricritical point a divergence is expected of heat capacity and compressibility, and, therefore, of such quantities as $1/c_{ij}$ [10]. However, nothing similar is observed in Figs 10 and 12. Nor is divergence observed in Fig. 8. Nevertheless, a specific anomaly, which manifests itself in the appearance of maximum jumps of elastic constants and of the coefficient of ultrasound absorption, is observed in the 0.3–0.4 T region of a

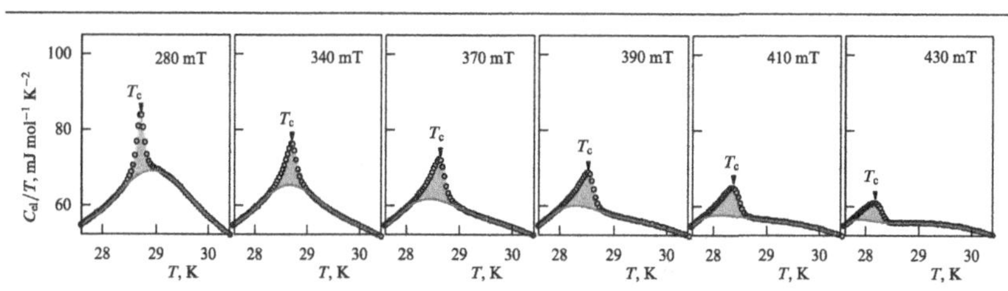

Figure 8. Heat capacity divided by the temperature at the phase transition in MnSi depending on the magnetic field [24]. According to the assumption of the authors of Ref. [24], a tricritical transition takes place in a magnetic field of 340 mT.

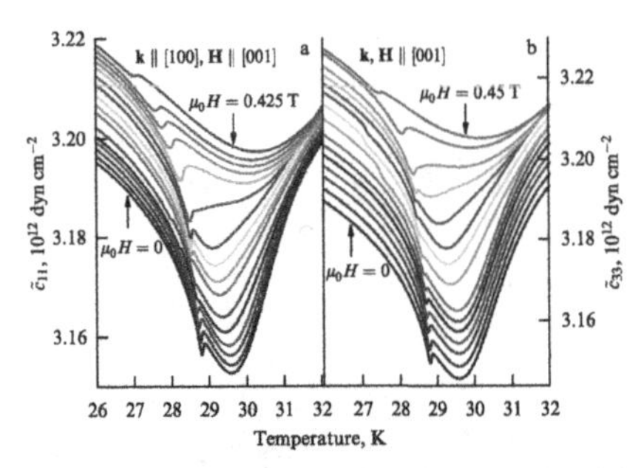

Figure 10. (Color online.) Temperature dependence of the elastic moduli $\tilde{c}_{11}$ and $\tilde{c}_{33}$ in the region of the phase transition in MnSi upon a variation in the magnetic field value. The tilde sign denotes the measured values corresponding to a cubic crystal with a tetragonal anisotropy induced by the magnetic field, in order to distinguish them from truly 'tetragonal' values [25]; **k** is the wave vector. The curves are shifted relative to each other along the ordinate axis.

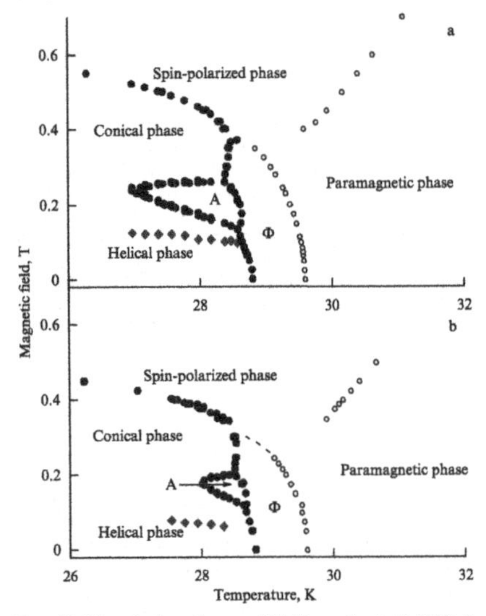

Figure 11. Magnetic phase diagram of MnSi according to Ref. [25]: A, skyrmion crystal; Φ, region of strong helical fluctuations. It can be seen that the domain of the existence of the skyrmion phase depends on the orientation of the sample in the magnetic field.

magnetic field. It is precisely here that the line of the minima of the elastic moduli corresponding to the smeared phase transition touches the line of phase transitions, which makes this region similar to an end critical point, at which the line of second-order phase transitions is joined with the line of the first-order transition [25] (see Fig. 11).

Additional information on the evolution of the magnetic phase transition in MnSi can be extracted from the data on

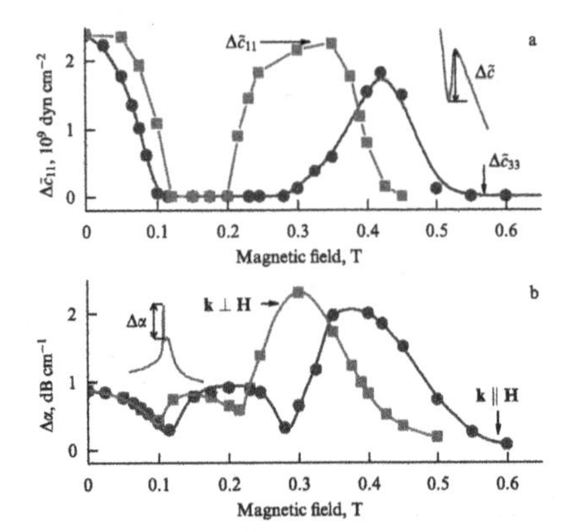

Figure 12. (Color online.) (a) Values of the jumps in elastic moduli, and (b) partial amplitudes of the attenuation coefficients at the phase transition in MnSi as functions of the magnetic field. The difference between the two systems of data given in figures (a) and (b) is connected with the difference among the magnitudes of the demagnetization factor [25].

the thermal expansion [26]. Some results of thermal expansion measurements are shown in Figs 13 and 14. It can be seen that in a magnetic field of 0.48 T the bulk anomalies disappear almost completely, which disagrees with the existence of a tricritical point in the magnetic field of 0.4 T in the case of the occurrence of a second-order phase transition for $\mu_B H > 0.4$ T.

Figure 15 displays the results of measurements of jumps in the sample lengths and in the heights of the peaks of the thermal expansion coefficient in the [100] direction at the phase transition in MnSi at various values of the magnetic field.

It can be seen that the anomaly of the coefficient of thermal expansion decreases with increasing magnetic field (see Figs 14, 15); a pronounced dip in the middle of the range corresponds to the region of the skyrmion phase (see Fig. 15). However, nothing indicates a tricritical behavior of the coefficient of thermal expansion. At the same time, the jump in the thermal expansion, decreasing to very low, but finite values (10^{-7}) with increasing magnetic field suddenly becomes zero. In reality, the jump is simply smeared, so that the measurement of its value becomes impossible. In Ref. [26], we concluded that the phase transition in MnSi is always a first-order transition, whose discontinuous picture is violated by heterophase fluctuations.

6. Phase diagram of MnSi at high pressures

In this section, we discuss the situation with the phase diagram of MnSi at high pressures. In papers [27, 28], Pfleiderer et al. stated, based on the measurements of magnetic susceptibility, that in the curve of the phase transition at a pressure of 1.2 GPa and a temperature of 12 K there is a tricritical point at which the continuous phase transition becomes a first-order transition. This idea found theoretical support [29]. And although the interpretation of the results of the measurements of the magnetic susceptibility

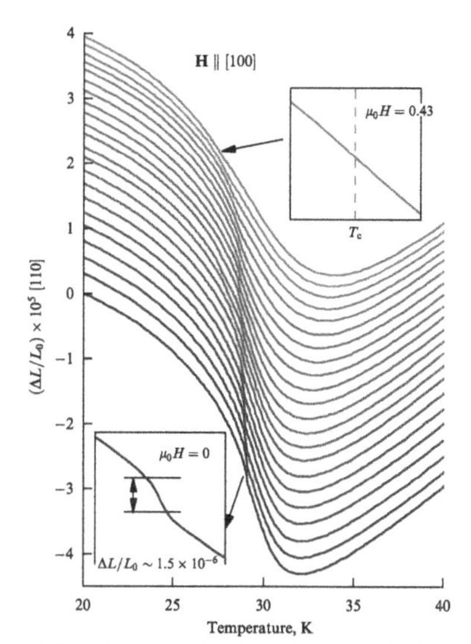

Figure 13. Linear thermal expansion of an MnSi crystal as a function of temperature at various values of a magnetic field. The curves are shifted relative to each other along the ordinate axis. It can be seen that the anomaly connected with the magnetic phase transition disappears in strong magnetic fields [26].

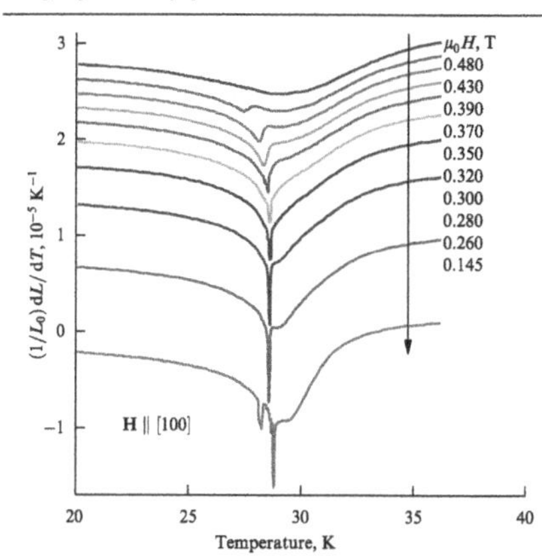

Figure 14. Temperature dependence of the linear thermal expansion coefficient of the MnSi crystal. The curves are shifted relative to each other along the ordinate axis. Degradation and disappearance of peaks of the coefficient of thermal expansion with an increase in the magnetic field are clearly visible. In the lower curve, a peak corresponding to the skyrmion phase is also noticeable [26].

of MnSi at high pressures was criticized (see review [1]), the 'tricritical idea' continued living. The measurements of bulky effects in the limit of low temperatures seemingly indicated

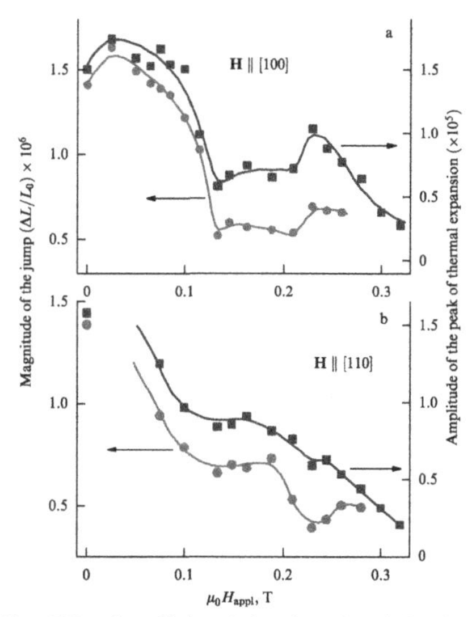

Figure 15. Dependence of the jumps in thermal expansion and values of the peaks of the thermal expansion coefficient of an MnSi crystal on the magnetic field. Both quantities decay with increasing magnetic field. The local anomaly in the middle of the range corresponds to the skyrmion phase. The jumps suddenly effectively decrease to zero in fields of 0.27–0.3 T (see the main text) [26].

the discontinuous nature of the volume change upon a phase transition in MnSi [30, 31]. In Ref. [32], this problem was analyzed in connection with the results of the measurements of the electrical resistance of MnSi. First of all, notice that the 'tricritical idea' was intensely popularized until the phase transition in MnSi was considered as a second-order one. At present, the situation, as we saw above, is fundamentally different, which, however, does not prevent us from examining the issue.

Figure 16 illustrates the dependence of the thermal expansion of MnSi on the temperature at atmospheric pressure. It can be seen that the weak first-order phase transition is hardly noticeable against the background of the extensive bulky anomaly. It is obvious that this transition could not be detected in the relatively rough experiments at high pressures [30, 31]. The authors of Refs [30, 31] apparently observed the bulky anomaly (Fig. 16), which, to a considerable extent, is localized as a result of 'freezing' thermal fluctuations at low temperatures and high pressures. This situation is illustrated in Fig. 17, where the isotherms of the electrical resistance of MnSi at different temperatures are plotted. It is clearly seen that the region of anomalous scattering of carriers on magnetic fluctuations shrinks with decreasing temperature and increasing pressure. At temperatures on the order of 2–5 K, the region of the anomalous scattering, one way or another connected with the volume anomaly [20], becomes too narrow to imitate the situation with the smeared first-order transition. However, these reasons do not force the authors of different concepts to

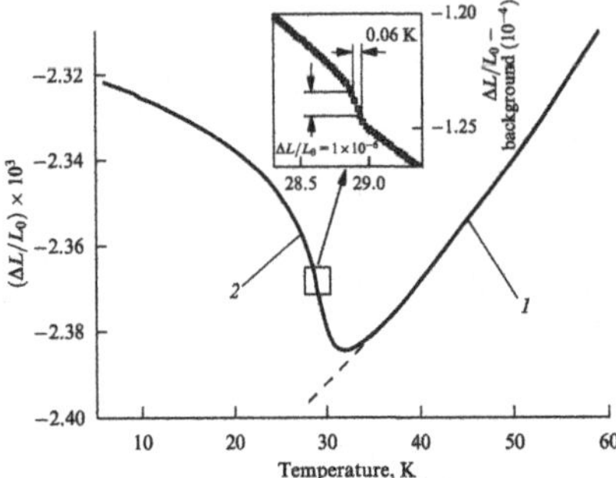

Figure 16. Linear thermal expansion of MnSi, which illustrates the relationship between the bulky anomaly and the first-order phase transition [32].

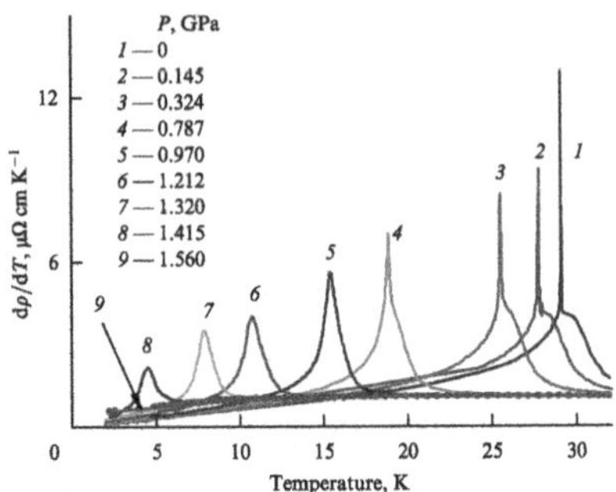

Figure 18. (Color online.) Temperature derivative of the resistivity $d\rho/dT$ at a phase transition in MnSi at high pressures.

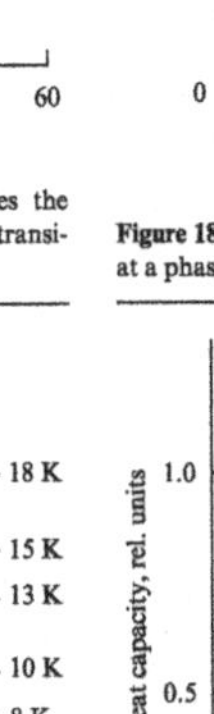

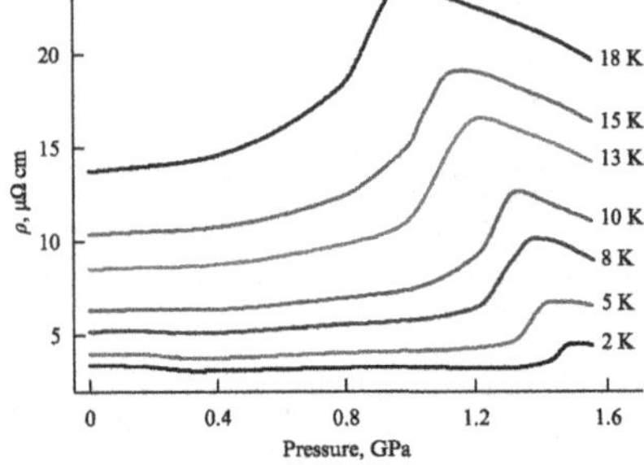

Figure 17. Isotherms of the electrical resistance of MnSi, which demonstrate the evolution of the fluctuation region in the vicinity of the phase transition [32].

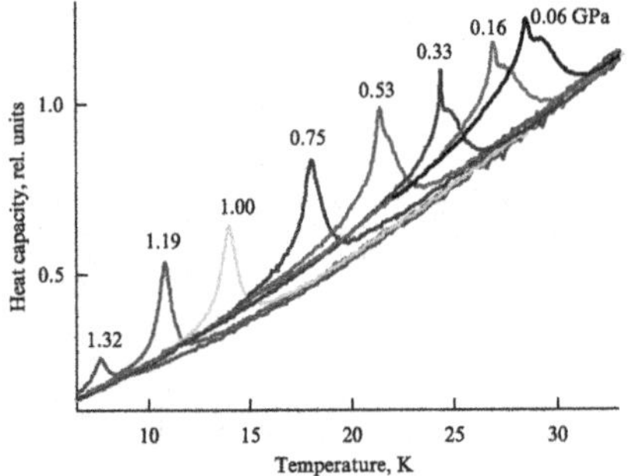

Figure 19. (Color online.) Heat capacity in the vicinity of the phase transition in MnSi at various pressures [33].

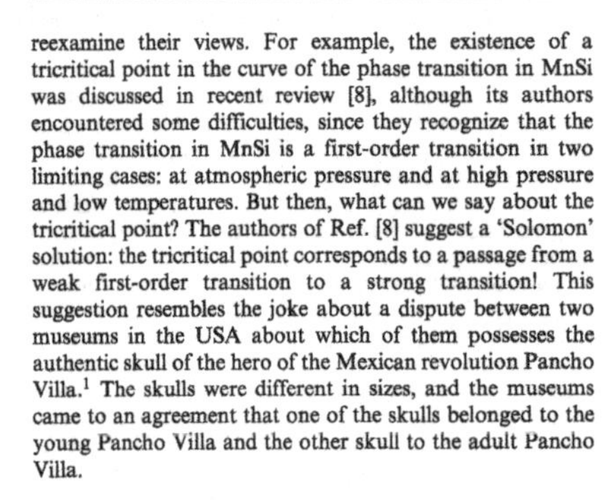

reexamine their views. For example, the existence of a tricritical point in the curve of the phase transition in MnSi was discussed in recent review [8], although its authors encountered some difficulties, since they recognize that the phase transition in MnSi is a first-order transition in two limiting cases: at atmospheric pressure and at high pressure and low temperatures. But then, what can we say about the tricritical point? The authors of Ref. [8] suggest a 'Solomon' solution: the tricritical point corresponds to a passage from a weak first-order transition to a strong transition! This suggestion resembles the joke about a dispute between two museums in the USA about which of them possesses the authentic skull of the hero of the Mexican revolution Pancho Villa.[1] The skulls were different in sizes, and the museums came to an agreement that one of the skulls belonged to the young Pancho Villa and the other skull to the adult Pancho Villa.

[1] Three years after Pancho Villa was assassinated, his grave was uncovered and his head was stolen.

Measurements of the electrical resistance [32] and heat capacity [33] at high pressures made it possible to draw specific conclusions about the phase diagram of MnSi (Figs 18, 19) (the temperature derivative of the electrical resistance in the case of phase transitions in magnetic metals behaves analogously to that of the heat capacity [34]). It can be asserted that the explicit signs of the first-order transition in MnSi [the sharp peak and the shoulder (see Fig. 3)] disappear with an increase in the pressure and a decrease in the temperature.

This may be connected with the suppression of thermal fluctuations, if we assume that the first-order transition in MnSi has a fluctuational origin. On the other hand, it cannot be ruled out that the phase transition is simply smeared at low temperatures and high pressures as a result of emerging the nonhydrostatic stresses.

Nevertheless, the results of work [32, 33] apparently indicate the absence of a strong first-order transition in MnSi as $T \to 0$. The phase diagram of MnSi proposed in Refs [32, 33] is shown in Fig. 20.

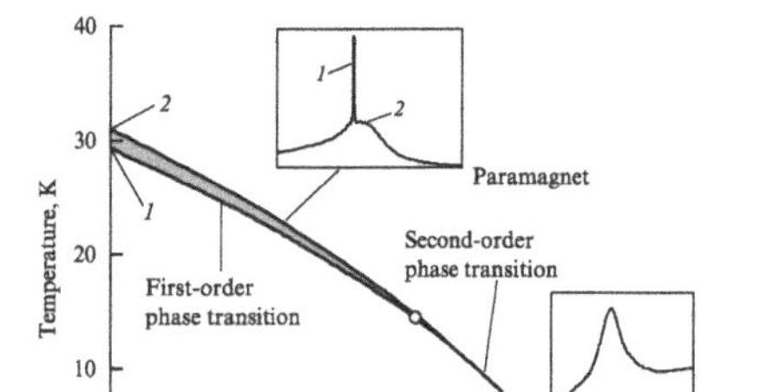

Figure 20. Supposed phase diagram of MnSi at high pressures. The gray region corresponds to the domain of strong helical fluctuations in the paramagnetic phase. The insets illustrate the evolution of the heat capacity and temperature derivative of resistivity $d\rho/dT$ depending on pressure. The circle at the beginning of the gray region can correspond to a tricritical point, if the phase transition in MnSi at high pressures is indeed continuous [32].

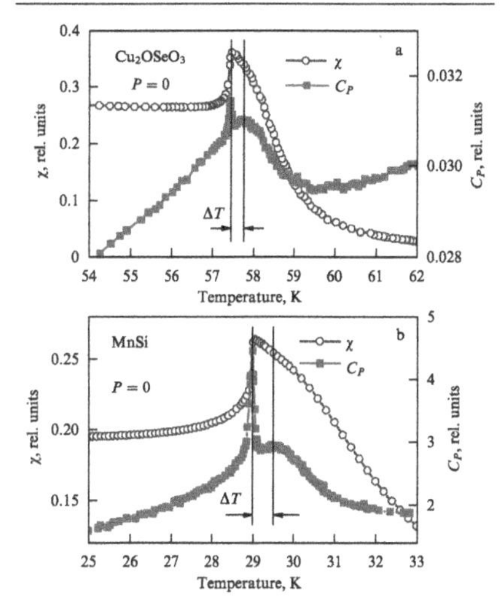

Figure 21. Magnetic susceptibility and heat capacity upon a phase transition in the chiral magnets (a) Cu_2OSeO_3, and (b) MnSi. The similarity between the behavior of the itinerant magnet and the magnet with localized spins is obvious [33].

7. Conclusion

This article is a kind of supplement to our earlier article [1] and does not pretend to an illumination of all achievements and difficult aspects of the problem. As a definite achievement, note the following: at present, it is already universally recognized that the magnetic phase transition in MnSi at atmospheric pressure and zero magnetic field is a first-order phase transition. However, this fact comes into conflict with early presentations of the shape of the phase diagram of MnSi at high pressures and low temperatures and the character of related quantum phenomena [27–29]. Progress in this field requires the development of new experimental techniques for studies at high pressures, which is a question for the future. Nor should we forget the uncommon phenomena that appear upon a phase transition in MnSi in strong magnetic fields [26]. And, finally, let us present Fig. 21 [33], which illustrates the close analogy between the itinerant and Heisenberg magnets, MnSi and Cu_2OSeO_3, respectively, with the Dzyaloshinskii–Moriya interaction, which, apparently, indicates the insignificant role of longitudinal spin fluctuations in the course of magnetic phase transitions [35].

Acknowledgment

This study was partially supported by the Russian Science Foundation (grant No. 17-12-01050).

References

1. Stishov S M, Petrova A E *Phys. Usp.* **54** 1117 (2011); *Usp. Fiz. Nauk* **181** 1157 (2011)
2. Thompson J D, Fisk Z, Lonzarich G G *Physica B* **161** 317 (1990)
3. Pfleiderer C, Julian S R, Lonzarich G G *Nature* **414** 427 (2001)
4. Doiron-Leyruad N et al. *Nature* **425** 595 (2003)
5. Pfleiderer C et al. *Nature* **427** 227 (2004)
6. Mühlbauer S et al. *Science* **323** 915 (2009)
7. Tonomura A et al. *Nano Lett.* **12** 1673 (2012)
8. Brando M et al. *Rev. Mod. Phys.* **88** 025006 (2016)
9. Pfleiderer C et al. *Phys. Rev. B* **55** 8330 (1997)
10. Landau L D, Lifshitz E M *Statistical Physics* Vol. 1 (New York: Pergamon Press, 1980); Translated from Russian: *Statisticheskaya Fizika* Vol. 1 (Moscow: Fizmatlit, 2002)
11. Rice O K *J. Chem. Phys.* **22** 1535 (1954)
12. Larkin A I, Pikin S A *Sov. Phys. JETP* **29** 891 (1969); *Zh. Eksp. Teor. Fiz* **56** 1664 (1969)
13. Halperin B I, Lubensky T C, Ma S *Phys. Rev. Lett.* **32** 292 (1974)
14. Bak P, Jensen M H *J. Phys. C* **13** L881 (1980)
15. Hansen P A, Ph.D. Thesis (Roskilde: Technical Univ. of Denmark, 1977); Report No. 360 (Roskilde: Risø National Laboratory, 1977); http://orbit.dtu.dk/files/53677871/ris_360.pdf
16. Janoschek M et al. *Phys. Rev. B* **87** 134407 (2013)
17. Brazovskii S A *Sov. Phys. JETP* **41** 85 (1975); *Zh. Eksp. Teor. Fiz* **68** 175 (1975)
18. Grigoriev S V et al. *Phys. Rev. B* **81** 144413 (2010)
19. Vollhardt D *Phys. Rev. Lett.* **78** 1307 (1997)
20. Stishov S M, Petrova A E *Phys. Rev. B* **94** 140406(R) (2016)
21. Bauer A et al. *Phys. Rev. B* **82** 064404 (2010)
22. Buhrandt S, Fritz L *Phys. Rev. B* **88** 195137 (2013)
23. Belemuk A M, Stishov S M *Phys. Rev. B* **95** 224433 (2017)
24. Bauer A, Garst M, Pfleiderer C *Phys. Rev. Lett.* **110** 177207 (2013)
25. Petrova A E, Stishov S M *Phys. Rev. B* **91** 214402 (2015)
26. Petrova A E, Stishov S M *Phys. Rev. B* **94** 020410(R) (2016)
27. Pfleiderer C, McMullan G J, Lonzarich G G *Physica B* **206–207** 847 (1995)
28. Pfleiderer C et al. *Phys. Rev. B* **55** 8330 (1997)
29. Belitz D, Kirkpatrick T R, Vojta T *Phys. Rev. Lett.* **82** 4707 (1999)
30. Miyake A et al. *J. Phys. Soc. Jpn.* **78** 044703 (2009)
31. Pfleiderer C et al. *Science* **316** 1871 (2007)
32. Petrova A E, Stishov S M *Phys. Rev. B* **86** 174407 (2012)
33. Sidorov V A et al. *Phys. Rev. B* **89** 100403(R) (2014)
34. Nabutovskii V M, Patashinskii A Z *Sov. Phys. Solid State* **10** 2462 (1969); *Fiz. Tverd. Tela* **10** 3121 (1968)
35. Stishov S M *Phys. Usp.* **59** 866 (2016); *Usp. Fiz. Nauk* **186** 953 (2016)

Printed in the United States
By Bookmasters